# MAKING
# **MATH**
# STICK

## Classroom Strategies that Support the Long-Term Understanding of Math Concepts

## DAVID COSTELLO

Pembroke Publishers Limited

*Special thanks to my wife, Marsha, and our two children, Faith and Cain, for their patience and support as I worked on this book.*

© 2021 Pembroke Publishers
538 Hood Road
Markham, Ontario, Canada L3R 3K9
www.pembrokepublishers.com

Library and Archives Canada Cataloguing in Publication

Title: Making math stick : classroom strategies that support the long-term understanding of math concepts / David Costello.

Names: Costello, David (Professional learning facilitator), author.

Identifiers: Canadiana (print) 20210149213 | Canadiana (ebook) 20210149264 | ISBN 9781551383507 (softcover) | ISBN 9781551389493 (PDF)

Subjects: LCSH: Mathematics—Study and teaching (Elementary)

Classification: LCC QA135.6 .C656 2021 | DDC 372.7/044—dc23

Editor: Margaret Hoogeveen
Cover Design: John Zehethofer
Typesetting: Jay Tee Graphics Ltd.

Printed and bound in Canada
9 8 7 6 5 4 3 2

# Contents

# Introduction

Teachers across classes, grade levels, and schools have been seeing the same problem. What we have here is a serious educational issue.

We have all experienced it: students learning math concepts only to forget them in the weeks and months that follow. The first time it happened may have seemed strange, but it now seems like a real problem. Students are not retaining what they learn. If we noticed just one or two students unable to recall previous learning, that might be an individual student issue. But that is not the case. Teachers across classes, grade levels, and schools have all been seeing the same problem. What we have here is a serious educational issue.

We know that teachers put their hearts into their work. We also know that school districts are revising math curricula and offering multiple professional learning opportunities to address the problem of students not retaining what they have learned. Unfortunately, we continue to see students unable to apply previous learning to novel situations. Our way to approach this has always been to stop and re-teach math concepts. In some cases, this re-teaching resembles cramming, which doesn't work in the long run and uses up valuable class time.

What is causing this lack of recall? It can't be the topic. Past generations of students have successfully learned mathematics. It can't be the time put in. Teachers are probably working harder than in times past. It could be the digital age we live in, which discourages people from putting information to memory. Few of us attempt to remember things because everything we need to know can be accessed easily via the Internet. Does this new inability to retain learnings have something to do with how we support our students?

Whatever the cause, I believe that we can address the issue effectively. I know because I have turned things around in my own classroom. The solution is not to work harder but to work smarter. We, as educators, can shift our paradigm from "teach, test, move on" to "teach, connect, apply," a move that can optimize student learning opportunities. This book is intended to provide you with an understanding of simple, manageable, and sustainable strategies you can use to work smarter.

Whether we are classroom teachers, coaches, administrators, coordinators, or consultants, we all have a lot on our plates. All too often we have tasks added to our plates while none are removed. I do not wish to add to your workload.

Instead, I hope to offer you a method for meeting the varying needs of your learners in a manageable and sustainable yet effective way. In the end, you will not have more work. You will simply be working differently.

The instruction and learning strategies I offer were developed based on evidence, in terms of both theory and classroom practice. They are straightforward—you could apply many of them in your classroom tomorrow. So that you can see these strategies at work in real classrooms, I have provided accounts from teachers and students sharing their experiences. You will also see work samples showing the strategies at work.

This book is meant to be one that can stay on your desk so that you can refer to it throughout the year. The strategies it offers are applicable to all mathematical concepts and classrooms. It is meant to support you in reaching whatever goals you establish for your particular students.

## Why We Need a Different Approach to Learning

So now to the classroom.

The following snapshots are taken from conversations with educators about their experiences in the classroom. As you read them, consider if you have experienced anything similar in your classroom.

These conversations are composite conversations typical of many I have had with teachers at all three levels. This book is filled with multiple examples of student work, as well as excerpts from conversations with teachers and students. Although they were contributed anonymously, all excerpts come from real conversations with real people.

### PRIMARY EXAMPLE: A TEACHER EXPLAINS

#### Forgetting how to estimate using a referent

I spent a considerable amount of time in the fall introducing estimation. More specifically, it was estimating with a referent. At first, we used concrete objects but then, over time, moved to pictures. It is now the winter and, when asked to estimate, my students will count the items and then record that as their estimate. I have tried, on many occasions, to encourage my students to use referents and have even shown a picture of a referent beside the items to be estimated. Still, no luck. I am at a loss and cannot spend any more time on this. Why were they able to demonstrate success with estimation using referents in the fall, but now it seems like they never experienced it? They had it so well but now say that they don't remember doing it.

### ELEMENTARY EXAMPLE: A TEACHER EXPLAINS

#### Forgetting how to work with prime and composite numbers

I was so impressed with my class for their work on prime and composite numbers. Students were able to identify which numbers were prime and which were composite when provided a list. Taking it further, students could identify the common factors for two given composite numbers and could list common multiples for given prime numbers. I would structure the lessons as follows: introduction, minilesson of a specific concept, and then practice. Students would be provided plenty of practice questions to apply their understanding of the concept in the minilesson. From there, we had a unit test, and all students performed very well. It has now been a few weeks, and I asked students to solve a few problems on prime and composite numbers. I was shocked! They were mixing up prime and composite numbers and were getting factors and multiples confused. What is going on?

### Forgetting how to solve equations

Solving equations is an important part of the Grade 8 math curriculum. So we spent a lot of time on this. At the end of the unit, students could solve equations involving a single variable. I assessed students and determined that they could isolate a variable whether it was added, subtracted, multiplied, or divided. Now, after March break, students cannot isolate the variable if it is multiplied or divided with another number. In fact, much of the class is saying this is hard and that they need step-by-step support when working on such equations. Where did all their success earlier in the year go? I thought that they understood what to do.

As you were reading the snapshots from the various classrooms, did you think back to times that you experienced similar unwelcome surprises? It can be quite perplexing and frustrating to realize that, although you spend a considerable amount of time on a concept, instructing and assessing, students seem to lose this learning as time goes on. You may have even communicated through report cards that students have demonstrated grade level understanding only to find out you need to review or re-teach a concept.

Looking back, what could you have done differently? You may be thinking, "I introduced the concept, provided students with multiple tasks to address and practice the concept, formatively assessed student thinking, addressed misconceptions, and then re-assessed to evaluate understanding. It seems that I have done all that I could do to address student learning. So why are all my efforts not working?"

The issue of poor retention is creating problems as students move from grade to grade. You have probably heard fellow teachers complain about having to re-teach concepts supposedly taught in the previous grade because students say that they either forget the concept or don't remember working on it the year before. While we sometimes accept this as normal, perhaps we should pause and think about why student failure to recall is becoming the norm. Are we okay with this being the norm? Should we be trying to address it? And how could we do so?

*Perhaps we should pause and think about why student failure to recall is becoming the norm. Are we okay with this being the norm?*

## Considering the One-and-Done Approach to Teaching Math

I would like to suggest, at this point, that the way that we structure the teaching of math may hold great potential for improving student recall. Consider the standard one-and-done organization of most math textbooks and year plans. Concepts are isolated into units and then each unit is segmented into smaller lessons. This is a great plan if you want to support short-term performance.

The following is a visual of a one-and-done year plan:

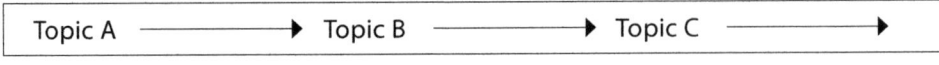

*Blocked practice*

Of significance is that the year plan is straightforward and linear. There is no returning to previously addressed mathematical concepts. Instead, the teacher covers Topic A, moves on to Topic B, and then to Topic C. Such an approach to instruction and learning is practice based on units of study. Such practice is referred to as blocked practice (Rohrer, 2009).

Consider how students would typically learn geometry in a Grade 8 classroom. The geometry unit has been compartmentalized into segments such as sketching a view of an object, drawing rotated objects, constructing objects from a certain perspective, identifying transformations, constructing tessellations, and identifying transformations in tessellations. During the first class, students receive a minilesson or exploratory opportunity on one of the segmented concepts. In the next class, they learn about a different segmented concept, in isolation. The pattern continues. While each of the segments plays a crucial part in student spatial reasoning, they are all treated individually. Students receive instruction during each lesson and are then assigned practice questions on the specific aspect of geometry addressed in the lesson.

In blocked practice, students are not required to consider which strategy to apply as this is already inherent in the questions and tasks assigned—all questions and tasks apply only to the single concept that was the focus of the lesson. A visual representation of this idea might be as follows.

**Lesson Topic A**
Practice questions:
1. Topic A    2. Topic A    3. Topic A    4. Topic A

**Lesson Topic B**
Practice questions:
1. Topic B    2. Topic B    3. Topic B    4. Topic B

**Lesson Topic C**
Practice questions:
1. Topic C    2. Topic C    3. Topic C    4. Topic C

*Question order typical of blocked practice*

> The one-and-done structure of most mathematical teaching does a significant disservice to students.

Notice that all practice questions relate directly to the focus of the lesson just taught. This removes the requirement that students select a strategy to apply to any given question or task. This drawback of the one-and-done structure of most mathematical teaching does a significant disservice to students. Missing are the cognitive processes that students would use in determining an appropriate strategy. These processes are crucial to mathematical understanding and problem solving.

Consider the potential impact of learning individual mathematical concepts in isolation through the course of a school year. While math is a network of concepts in which there is much interconnectivity, the one-and-done school year is comprised of segmented learning. Instead of working through the intricate tapestry of mathematics, concepts are compartmentalized.

While the logic behind such planning may be that teachers can ensure that they cover all required grade level content and give each unit time, we have to ask: Is it best for student learning? Are these units of study building blocks or stumbling blocks? When mathematics is chunked up into isolated concepts, how much time do students spend making connections to previous learnings? I assert that we do an injustice to mathematics learning when we fail to introduce students to these connections.

And what is the result of teaching math concepts in isolation? Unfortunately, we experience incidents like those described in the classroom snapshots above. We observe students demonstrating understanding of the concept when that concept is the instructional focus, but then those same students being unable to replicate their understanding later in the year. We must be cognizant that just because a student can do a problem one time does not mean that this student will be able to recognize the type of problem and remember the appropriate strategy for solving it later. Almost no student masters something new after one or two encounters with it.

Within our education system there are many established structures that promote short-term performance instead of long-term learning. Such structures have been entrenched in our system for decades (year plans and textbooks being just two of these) and seem to focus on dispensing information instead of supporting students in recalling and applying that information. To help our students, we need to recognize and address the limitations of these structures. We need to adjust our instruction to nurture long-term learning. Our goal, as an education system, should be to support students in acquiring knowledge and skills that they can retrieve and apply not only in the short term but also in the long term.

## What We Mean by Learning

Let us step back for a moment and think about how students learn. How do they make meaning? How do they use newfound knowledge to strengthen their understanding?

From the perspective of cognitive science, learning is the acquisition of knowledge and skills and having these readily available from memory to support future meaning-making opportunities. More specifically, there are three stages of learning: encoding, consolidation, and retrieval (Brown, Roediger III, & McDaniel, 2014).

- **Encoding** means getting knowledge into our heads.
- **Consolidation** means assigning meaning to that knowledge, which gets it into our memory. It is through this process that connections are established between recently encountered knowledge and knowledge already stored in long-term memory.
- **Retrieval** means getting knowledge out of our heads. Retrieval is the process of reaching back and bringing something we previously learned into mind.

When thinking about how to teach, it is important to consider all three stages of learning. Each of the stages plays a considerable role in the learning process. Without all three stages, learning becomes far more challenging and not as effective.

Let's look at the three stages of learning in relation to additive fact strategy and the make-ten strategy.

- **The encoding stage.** Students, whether through explicit instruction or exploratory activities, realize that they can apply the make-ten strategy to fact learning. Through this process, students encode the make-ten strategy into their short-term memory.
- **The consolidation stage.** Not until students consolidate their new learning with other relevant and interconnected previous learning that they move the make-ten strategy from short-term memory to long-term memory. An example of this would be when students make connections between the make-ten strategy, sums to 10, and part-part-whole. By connecting these three strategies, students can make ten and then add on the remaining part of the addend.
- **The retrieval stage.** When students recall and apply this make-ten strategy, they are in the retrieval stage of learning. It is during retrieval that they look for the knowledge in their memory and apply it to a given situation.

> **The more we use a strategy, the more accessible it becomes. The less we use it, the more difficult it is to access.**

It is important to note that our memory stores information based on relevance and use. The more we use a strategy, the more accessible it becomes. The less we use it, the more difficult it is to access. Hence the necessity to retrieve new information from memory multiple times over a period of time. The more students retrieve the new information, the more securely it sits in long-term memory. This retrieval could take the form of drills, practice, or other challenges. One caveat is that student must understand the strategy before moving to practice.

By focusing on the three-part learning process of encoding, consolidation, and retrieval, you can develop an instructional approach highlighting learning as opposed to memorization. By working through the three stages of learning, students can strengthen their use of strategies as well as their strategy-selection skills. With multiple retrieval opportunities, students' recall improves. Students thereby develop automaticity with the facts. Here we have the difference between memorizing (putting isolated facts into memory) and developing memory (by understanding and repeatedly retrieving).

## Why Encoding Is Not Enough

Learning works best when it encompasses all three stages of learning (encoding, consolidation, and retrieval). Which stage do teachers tend to focus on in class?

The following are some comments I have received from teachers and students in response to this question.

---

**THOUGHTS FROM TEACHERS**

- "It's important to recognize all the grade level content that students must master during the year. We have to ensure that they have opportunities to understand the concept."
- "I have outcomes to teach and it's important that students are exposed to this content. They have to see it, apply it, and understand it."
- "I have a lot of students with learning needs in my class, and it's up to me to support all of them. I don't have time to keep coming back to concepts.

---

I just make sure that I do a concept justice during its time in my yearly plan."

- "Differentiation. I need to plan instruction so that all my students can access the concept that is the focus of the lesson."
- "I have four classes of Grade 8 math, and I need to make sure that each class gets through the curriculum. There is only so much time, and it's important that I don't create any gaps for the next grade level."

### THOUGHTS FROM STUDENTS

- "I feel like I'm a sponge. I don't like math so much, but I know it's necessary to understand the strategies that we are being taught."
- "It's about showing the teacher that we know how to do what she is teaching us."
- "I have six classes some days and I have lots of stuff to learn. The teacher goes over the topic and then gives us time to practice it. Same thing every day."

If these opinions are representative, then we can safely say that the classroom focuses on the encoding stage of learning. Why encoding? A few factors may explain why this is the case.

- First, teachers are provided with a curriculum to which they must adhere. They are responsible for ensuring students receive instruction and experiences to cover all grade level expectations up to set standards. Consequently, they feel pressured to move through the curriculum and standards at a brisk pace to ensure that they cover all necessary content. This pace does not allow for delving deeply into concepts.
- Second, classroom assessments most often measure students' ability to demonstrate understanding of the content that has just been covered. This assesses student knowledge and understanding on a short-term basis. It can be thought of as the path of least resistance because it does not require the student to retrieve learning from long-term memory. This is likely the way that we were taught as students (Kindergarten through post-secondary) and we assume that this is how you learn.
- Third, students are required to take provincial or state standardized assessments to evaluate their level of mastery of mathematical content for their grade level. To ensure students do well on these assessments, teachers teach to the test. When we teach to the test, our focus tends to be on helping students respond correctly rather than helping them understand. It is about getting material into students' heads, in other words, the encoding stage.
- Fourth, teachers may not be aware of the benefits of emphasizing all three stages of learning.

**When we teach to the test, we tend to focus on helping students respond correctly rather than helping them understand.**

When focusing on the encoding stage of learning, attention is paid to short-term performance.

In the usual classroom, the teacher instructs students (or helps them to explore) a specific concept. The teacher immediately provides practice questions and tasks. At the completion of a unit, the teacher assesses student understanding of the concept, possibly with a test. Many students have little to no difficulty with the assigned tasks and tests. Unfortunately, both the teacher and these students conclude that the students have mastered the relevant math concept. This belief is a fallacy, however, because students have not learned to solve the questions or tasks without knowing the strategy in advance. This fallacy is referred to as illusion of mastery (also illusion of fluency and illusion of knowing) (Agarwal & Bain, 2019; Brown, Roediger III, & McDaniel, 2014).

Illusion of mastery is problematic for both the teacher and students. It results from misunderstanding which learning strategies ensure long-term performance. When all they can see is short-term performance, both teachers and students assume that comfort using a skill or strategy in the short term equates with effectively using it in the long term. Teachers and students both conclude that students who have demonstrated understanding of a concept once will be able to apply it moving forward. This fallacy in perspective results in teachers and students favouring learning strategies that promote short-term retention (Nazari & Ebersbach, 2019) instead of learning strategies that are effective for both the short term *and* long term.

> When all they can see is short-term performance, both teachers and students assume that comfort using a skill or strategy in the short term equates with effectively using it in the long term.

When we focus solely on encoding, we are not providing students with opportunities to consolidate new learning with previous learning, retrieve learning from previous experiences, or work through the important step of strategy selection in fact learning. While students would perform well on immediate assessments, they may not perform well in the long term.

Let's consider how we might incorporate consolidation and retrieval into the learning process. During consolidation, students would layer their previous learning with the newer concepts and begin to process similarities and differences between the two. This not only strengthens the newer learning but also reconstructs the previous learning. To facilitate retrieval capacity, students would be required to access the newer learning repeatedly, thereby adding new layers of meaning.

Let's consider the double-double multiplicative strategy. Here is an example of it:

The task: Solve for $7 \times 4$.
– Use $7 \times 2$ to find the product of 14.
– Add 14 and 14 (double 14) to find the sum of 28.
Therefore, $7 \times 4 = 28$.

If you focus teaching and practice solely on teaching the double-double strategy, you are putting all your efforts into the encoding stage. How could you expand your teaching of the double-double strategy to include consolidation and retrieval?

- For **consolidation**, students might process this strategy in terms of similarities and differences between the double strategy, clock facts, and nifty-nine, thereby strengthening all four strategies.
- To provide **retrieval** practice, you could provide students with opportunities to practice selecting appropriate strategies to solve problems. For example, you could provide students with multiple problems and have them circle those that the double strategy would be effective in solving, place a square around those that the clock-facts strategy would be effective in solving, and a triangle around those that the nifty-nine strategy would be effective in solving. Such

retrieval practice provides students with the opportunity to make more connections to past learnings.

## Only Effortful Learning Lasts

Students enjoying an illusion of mastery will not retain their learnings. Solving a problem when the strategy is given is relatively easy. If our goal is to help students to apply knowledge from memory to make sense of future problems, then we must make the learning effortful. Learning must be developed over time through meaningful, purposeful practice—and it must be hard!

We can make learning effortful by introducing some challenge to students, in other words, desirable difficulty (Roediger III & Karpicke, 2006). Desirable difficulty, simply stated, is the process of retrieving forgotten knowledge. It is the struggle to interrupt the forgetting that naturally occurs after initial encoding. After students forget a piece of knowledge, they must work to retrieve the knowledge from their memories. Even when students do not think they have the memory, it is there, hidden deep inside.

Desirable difficulty can be thought of as productive struggle. Productive struggle is an important component of mathematics, especially in relation to making meaning. Productive struggle leads to understanding and it makes the achievement of learning goals possible. Productive struggle occurs when the student searches for and finds knowledge that is not immediately available. The search for forgotten memories is the struggle; the struggle becomes productive when the student regains memory of a strategy that can be applied to a task.

For learning, the desirable difficulty is the process of retrieving knowledge and understanding from long-term memory. When we recognize and understand the role of desirable difficulty in the learning process, we can take action to strengthen student learning. When we communicate this idea with students, they will appreciate that the difficulty is something they need to work through and that it helps them learn.

## How Retrieval Boosts Learning

Many people think of retrieval, the third stage of learning, as a knowledge check, a dipstick they can use to see if they know something (Brown, Roediger III, & McDaniel, 2014). A common misconception is a belief that if students can recall knowledge once, the learning is complete. Research reveals, however, that the practice of retrieving multiple times helps individuals learn—the more times we retrieve something, the stronger our memory of it becomes (Agarwal & Bain, 2019; Brown, Roediger III, & McDaniel, 2014; Roediger III & Karpicke, 2018).

Let's consider the practice of retrieval as a form of exercise. The more that we retrieve, the more defined the pathways to memories become (Roediger III & Karpicke, 2006). While the pathways may take significant cognitive demand initially, over time these pathways are defined and strengthened by an increasing number of retrieval cues. That is the benefit of retrieval in terms of creating pathways to previous learning.

Now, let's consider retrieval in terms of the actual learning. Each time a memory is retrieved, the individual must think about it once again, re-construct it in terms of the previous knowledge and current experience, and then re-consolidate it

(Brown, Roediger III, & McDaniel, 2014). Through this process of re-construction and re-consolidation, the memory is strengthened through multiple layers of involvement—every time a person interacts with a memory, they bring new understanding and experiences to add further structure to the memory.

The brain organizes our memories more efficiently every time they are retrieved. By strengthening pathways to memories and increasing the retrieval cues, the brain naturally makes these memories more accessible. Through re-construction and re-consolidation, memories are layered and contextualized by multiple interactions.

We know that retrieval produces learning. We can go further to argue that retrieval is, in fact, a more powerful learning activity than that of encoding (Karpicke & Blunt, 2011).

> By strengthening pathways to memories and increasing the retrieval cues, the brain naturally makes these memories more accessible.

## How to Change Your Teaching Practice

Changing the focus of learning in the classroom to incorporate aspects of retrieval will provide the desirable difficulty needed to strengthen retrieval cues and add layers of memory. For this book, I will highlight three ways to increase retrieval opportunities in your classroom: spaced practice, mixed practice, and feedback. Let's have a look at all three individually, and then examine how they can work together.

When thinking of spaced practice, think about time. To space retrieval means literally to spread lessons and retrieval opportunities over a period of time instead of squeezing them together (Agarwal & Bain, 2019; Brown, Roediger III, & McDaniel, 2014). When students return to content often, they interrupt the forgetting process before the initial memory gets completely lost. The longer the period of time before retrieval of a memory, though, the harder will be the struggle to retrieve it. The difficult act of retrieval then strengthens the memory by re-constructing and re-consolidating it while strengthening the pathways that lead to it.

Mixed practice (also called interleaved practice) is another approach to retrieval. To interleave means to mix up problems so that consecutive problems cannot be solved by the same strategy (Agarwal & Bain, 2019; Brown, Roediger III, & McDaniel, 2014). When problem types come in random order, a student must first think about the problem and determine the necessary strategy before applying it. Compare this approach to the traditional presentation of practice questions in textbooks. Here, the strategies necessary to solve the questions are all the same and are all the focus of the preceding lesson. Mixed practice places greater cognitive demand on students (Agarwal & Bain, 2019; Brown, Roediger III, & McDaniel, 2014). The concepts addressed in the questions must not be similar, so that students undergo the difficulty of selecting a strategy and then applying it.

Both spacing and mixing practice can greatly increase retrieval opportunities. You should note that these practices place much more cognitive demand on students. Students cannot just sail through a set of similar problems. Instead, they have to retrieve memories about multiple strategies they have encountered and decide which ones would work best for each problem. They may have to retrieve a memory from earlier in the year, one that they haven't applied in quite some time. This spacing and mixing thereby crafts a desirable difficulty of productive struggle.

The following is a conceptual example illustrating the difference between blocked practice and a spaced and mixed practice. In the example, the letters A, B, C, and D each represents a particular concept. Let's assume that you have taught all four concepts, and that you have scheduled time for four practice sessions.

|  | Practice Session 1 | Practice Session 2 | Practice Session 3 | Practice Session 4 |
|---|---|---|---|---|
| Blocked Practice | AAAA | BBBB | CCCC | DDDD |
| Spaced and Mixed Practice | ABCD | BDCA | CBDA | DABC |

*Comparing the order of questions in blocked practice versus spaced and mixed practice*

Notice that no additional work demand is placed on your shoulders and no additional time is required. Your students are still assigned four practice problems for each concept, and they are engaged in the same number of practice sessions. What is different is that for each session, your students will address one question per concept, and the concepts are spread out over four sessions.

In spaced and mixed practice sessions like this, short-term performance may be hindered initially because of the larger cognitive demand of this type of practice—students may need to work hard to recall and select the right strategy and they might make mistakes. Over time, however, student comfort with this approach increases and long-term learning is enhanced. Before long, students gain aptitude in effectively selecting and applying strategies and concepts.

A further benefit of spaced and mixed retrieval practice is that learners themselves identify any gaps in their learning (Roediger III & Karpicke, 2006). This ability to self-identify gaps in learning is a form of metacognition—it's about being aware of what you know and what you don't know. Through spaced and mixed retrieval, students' evaluations of what they know and don't know becomes more accurate. This understanding can only improve with repeated opportunities for retrieval. The more that students retrieve previous learning, the more they self-assess their thinking and identify opportunities for growth.

**The more that students retrieve previous learning, the more they self-assess their thinking and identify opportunities for growth.**

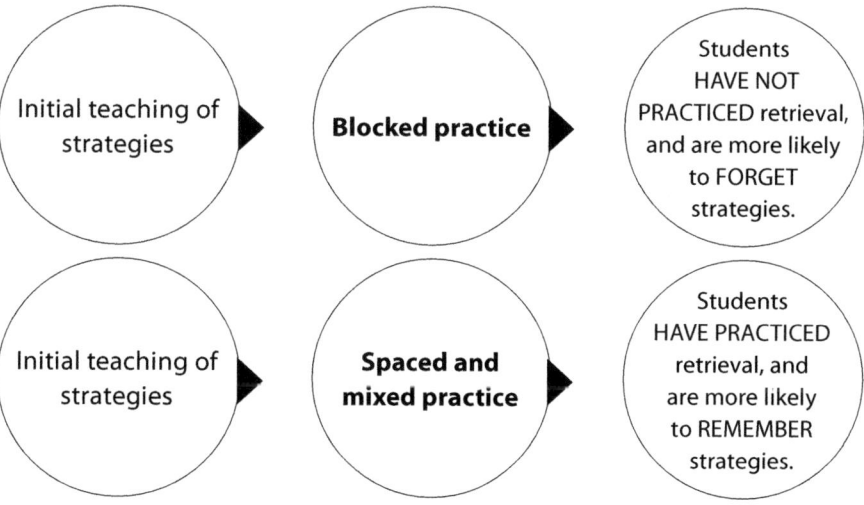

*Comparing retrieval opportunities in blocked versus spaced and mixed practice*

While students have opportunities for self-assessment through ongoing retrieval practice, it is important for you to provide students with feedback so that they can recognize whether their approaches to solving the task are correct or incorrect (Agarwal & Bain, 2019; Brown, Roediger III, & McDaniel, 2014). Your feedback can support student metacognition as they navigate mathematical concepts. It can reinforce or correct their awareness of their learning, so that they can then take any necessary next steps. Whether you provide feedback immediately after the task is completed or after a delay appears to be immaterial (Agarwal & Bain, 2019). What is important is that feedback is given at some point because it makes learning durable.

Spaced and mixed retrieval practice accompanied by feedback can benefit all students. Think about that: *all* students. This refers to our students who are English language learners, students who are having difficulty with mathematics, students who have demonstrated success with mathematics, and students with individualized programs. Regardless of your students' learning needs, retrieval practice will make their learning durable and will provide them with a personalized next step in their learning journeys. You will be supporting your students in becoming independent learners as opposed to passive learners.

> Spaced and mixed retrieval practice accompanied by feedback can benefit all students. Think about that: *all* students.

## In This Book

This introduction has identified a serious and widespread lack of retention capabilities among students in math class. The means to address this problem lies in individual educators' hands. All that is required is an appreciation of the three stages of learning (encoding, consolidating, and retrieval) and a willingness to try a different approach to instructional practice, one that places a stronger focus on retrieval practice.

In the chapters that follow, you will find manageable and sustainable strategies you can use to support your students in retrieving and strengthening previous learning. Your students can learn to increase their ability to recall learning from earlier in the term, year, and previous grades, and then to apply this learning in novel situations. Part 1 and Part 2 offer two types of strategy: one for the student and one for the teacher.

Part 1 offers efficient and effective learning strategies designed to help students build their retention capabilities. You can incorporate them easily into your pre-established classroom routines, via assignments, homework, reflection activities, and so on. These strategies range from self-quizzing to dual coding. After you practice the strategies together as a class, students can sort out which ones work best for them to use individually.

Part 2 offers instructional strategies you can use to design a personal teaching approach that helps students build their retention capabilities. You can support students in not only understanding mathematical concepts but also recalling and applying these to novel situations. These strategies range from designing a year plan to making better use of the exit ticket.

The strategies you encounter on these pages will make your teaching more effective. By making use of them, you will help your students remember what you have taught them. Thereby, you will make them better students and help prepare them for future success.

# 1

# Learning Strategies That Build Retention

**You have the power to move your classroom from being a place where students absorb information to a place where students are fully engaged in the process of retaining what they learn.**

As we explored in the introduction to this book, learning has too often been thought of as encoding—the initial step of getting information into students' heads. For learning to be effective, though, encoding is not enough. Learning is a process that involves encoding but also consolidation and retrieval.

So how do you change your classroom to be a place where all three stages of learning take place? Half of the answer lies in instructional strategies, which you can read about in Part 2. First, though, I'd like you to consider the learning strategies in Part 1, which constitute the other half of the answer. These strategies are ones that your students can use themselves to become more aware of and improve their own learning process. They are designed for students to use independently to consolidate and retain the learnings you have introduced.

You have the power to move your classroom from being a place where students absorb information to a place where students are fully engaged in the process of retaining what they learn. You are not alone in the task of helping your students learn—engage your students! You and your students can try out the strategies as a class, and then individual students can figure out which ones best help them retain learnings. This is metacognition at the classroom scale.

## Accurate Self-Knowledge Assists the Learning Process

Independent learners are those who can apply what they learn to new situations and who can recognize the knowledge they have and the areas that need additional focus. It is not enough that students passively respond to your hints to help

Students must be cognizant of the significant role they have in their own learning. You can help increase their awareness by talking about the learning process openly.

them to recall previous learning and to strengthen their understanding. Students must be cognizant of the significant role they have in their own learning. You can help increase their awareness by talking about the learning process openly.

As we learned in the Introduction, many teachers hear the terms *forget* and *forgot* in the classroom. Usually forgetting a learning is viewed by teacher and student alike with frustration. I propose that we embrace forgetting as a natural occurrence within the learning process. Acknowledging forgetting as something that is not a negative but a necessity is what Henry L. Roediger III and Jeffrey D. Karpicke (2006) termed a *desirable difficulty*. Learning is strengthened when students put effort into retrieving forgotten knowledge from memory. When students recall information, they boost their learning as the retrieval of previous learning strengthens the pathways to this learning and reconsolidates the memory. It is this effort that we want to acknowledge and to have students accept as necessary components of learning.

You can help students attain more comfort with forgetting as a necessary step in learning simply by accepting everyday examples of forgetting as opportunities. For students to accept the desirable difficulty of forgetting, however, they may have to refine their understanding of themselves as learners. Many students are under the assumption, as false as it is, that if something comes easily to them, then they understand it and will remember it. This is not the case. Remembering easily may indicate a likelihood of strong short-term performance but not strong long-term performance. This misconception, referred to as illusion of mastery (Agarwal & Bain, 2019; Brown, Roediger III, & McDaniel, 2014), is a fallacy that you can help students grasp.

Consider the following examples of student responses to queries about their performance. These students are representative of those experiencing an illusion of mastery. All three were shocked at their long-term performance on classroom assessments.

### PRIMARY EXAMPLE: A STUDENT EXPLAINS

#### Illusion of mastery of word problems

We spent a lot of time on word problems in the fall. I mean a lot of time. We would add or subtract numbers and then be given word problems to do. Some days, the teacher would give us, like, five or ten addition problems and about ten minutes to solve them. We would then check them as a group. Then the teacher would give us, like, five word problems to solve. I always got mine right. Then other days we would do the same, only subtract.

But it's not fair what happened last week. We were working on finding the perimeter of shapes. Then, on Friday, the teacher gave us three word problems to solve. All the problems were about subtraction and money. I thought I had them all right, but when the teacher went over the answers with everyone, I had them all wrong! I was confused. We were working on perimeter all week—how was I supposed to know that the problems were about something else?! I thought that I knew how to find the answers for word problems, but I guess I don't.

## Illusion of mastery of place value with decimal numbers

At the start of the year, we did place value with whole numbers. Then we got to work on decimals. It was the same as whole numbers, only we were using decimals. I did great with expanded form, standard form, and writing numbers in words. But that was at the beginning of the year.

Before the second report card, my teacher gave us some questions on decimals. I didn't know what I was doing. I couldn't remember to put "and" in for the decimal, and I forgot the place value names of the places to the right of the decimal.

I usually get all questions right in class, but I didn't this time. The teacher should review the concept with everyone before giving us questions.

## Illusion of mastery of order of operations

Doing BEDMAS was fun in class. I remembered doing some of it in Grade 6, but this year we used it more. At the end of the year, we had a little end-of-year test where the teacher gave us questions on all the work we did during the year. I had every one of the BEDMAS questions wrong. I did all the operations from left to right instead of doing brackets first, then exponents, then multiplication and division, and then addition and subtraction.

It's kinda my fault because my teacher gave us a review to be done for homework, but I didn't do it because I didn't think I had to. I got all A's during the year, so I didn't think I needed to do the review work at home.

It really doesn't make sense. How come I did so well doing BEDMAS in the first of the year but not now?

It is fair to say that the students represented in the above examples realized that their assessment of their learning was inaccurate. Before being disillusioned, the students did not have an accurate awareness of what they knew and what they needed to practice. Metacognition—the ability to self-identify gaps in learning—is thinking about one's thinking process. It's about knowing what you know and knowing what you don't know. We want students to have strong metacognitive skills so that they are not subject to illusion of mastery and so they can identify reasonable personal learning goals.

Think about the following situation in relation to your own classroom experiences. Have your own students suffered similarly from illusion of mastery?

## Illusion of mastery of simple and mixed fractions

A teacher assigned a test in his elementary math class. Students complete the test individually and without teacher support. Let's consider two students:

Student 1 confidently walks up to the teacher after completing the test and passes it in. The teacher asks this student if he reviewed his work. The student replies "yes" and comments on how easy the test was.

Student 2 asks multiple questions during the test. Each time, the teacher tells the student to read the problem and to think about strategies that could be applied to reach a solution. When this student hands in the test, the teacher asks if she reviewed her work before passing it in. The student replies "yes" but says she is unsure of her work.

In the next class, student 1 is shocked when he receives the test and sees that he didn't do very well at all. He handled the simple fractions well but not the mixed fractions.

Student 2 is happily surprised when she receives the test and sees that she answered most of the questions correctly.

It is fair to say that *both* students could benefit by strengthening their metacognition. Student 1 was under the illusion of mastery, while student 2 didn't have a great understanding of her knowledge.

Neither student accurately identified their knowledge or gaps in their knowledge.

What might have happened had those students had a better understanding of the status of their knowledge before taking the test? Would they have been more inspired to review and prepare? Would they have been better able to zero in on the gaps in their knowledge?

It is all too common that students do not understand how well they are doing—whether it is positive or not. Ideally, they should not have to rely on their teacher to tell them if their work is correct or incorrect. Lack of self-awareness means that students cannot self-monitor as they work through a problem and refine their approach as necessary. Nor will they know if and where to focus their study efforts. The good news is that metacognition can be strengthened through practice.

## Forgetting and Remembering

I started this discussion by reviewing the broader meaning of learning to include encoding, consolidation, and retrieval. By keeping all three stages of learning top of mind, we can provide students with opportunities to exercise their metacognition thereby supporting them as independent learners. By doing this, students will strengthen their previous learning and be able to transfer their learning to novel situations—meaning they can apply previous learning to solve current and future problems.

You will do your students a great service by helping them understand that short-term performance is not a good indicator of long-term retention of learning. Instead, learning is a long process that involves forgetting and remembering, and then forgetting and remembering again. Only then can students be sure that they will retain a learning and be able to apply it to novel situations.

The learning strategies in Part 1 are designed to support students in strengthening their ability to recall knowledge. By recalling knowledge repeatedly, they consolidate it and are better able to retrieve it in future. By introducing your

**Learning is a long process that involves forgetting and remembering, and then forgetting and remembering again.**

students to these learning strategies, you will enable them to solve problems more effectively and efficiently, as their previous learning will be solidified and the pathways to this learning will be strengthened.

## How to Apply These Strategies

While some of the following strategies may seem familiar (e.g., practice with flashcards), our focus is not on the strategy itself but instead on *how* students can apply the strategy to strengthen their learning. All the strategies and corresponding examples constitute opportunities for retrieval that can build student learning and metacognition. By spacing and mixing retrieval opportunities, students will be able to strengthen both previous learning and pathways to it. With practice, students will be better able to apply their learning to current and future problems.

Before we begin to explore strategies and examples, I want to reiterate the importance of effort in learning. When students are engaged in learning strategies and encounter stumbling blocks that require effort to overcome, they experience productive struggle. It is this productive struggle that strengthens student learning. After successfully retrieving the memory, both the knowledge and ability to retrieve it from long-term memory will be bolstered. Communicating this phenomenon with students is key: the process works best when students are aware of the process they are engaged in.

While the learning strategies provided in Part 1 will support students in retrieving previous learning and applying it to new contexts, how these strategies are utilized cannot be overlooked. Therefore, as I share the strategies and examples, I will provide suggestions as to how and when students would engage with the strategies both in and out of the classroom. It is important that students have opportunities to apply these strategies over a period of time, whether it be the term, semester, or school year.

After students apply a strategy, be sure to give them an opportunity for feedback. This feedback can be given by you or by themselves while looking over their responses. Feedback does not have to be immediate, but it does have to be timely. Students need to know quickly whether they are correct or not so that they can adjust their thinking to strengthen their ability to self-assess what they know and don't know.

The following table highlights the student learning strategies explored in the chapters in Part 1. On their own, the student learning strategies could assist students in making meaning of concepts, that is, during the encoding stage. However, timing is critical—it is *when* students use these strategies, that is, during the consolidation and retrieval stages, that will most strengthen their understanding and strengthen their ability to apply previous learning in new situations. You can help your students use these strategies to interrupt the forgetting process, overcome their illusion of mastery, and address inaccurate assessments of their understanding.

The process of building retention capability works best when students are aware of the process they are engaged in.

| Student Learning Strategies | | |
|---|---|---|
| **Strategies** | **In the Classroom** | **Examples Explored** |
| **Chapter 1:**<br>**Learning through**<br>**Self-Assessment** | Help your students to use objective tools to gain a realistic perspective of what they don't know yet and consolidate what they do know. | Students<br>• create their own **self-quizzes**<br>• use **flashcards** for mixed and spaced review<br>• **self-monitor** |
| **Chapter 2:**<br>**Building a Network of**<br>**Memory** | Help your students elaborate their knowledge of a concept by adding layers of meaning to a memory. | Students<br>• create and answer **how and why** questions<br>• **explain** their thinking<br>• use **dual coding**<br>• **compare and contrast** concepts<br>• generate **concrete examples** of abstract ideas<br>• **make connections** to prior knowledge |
| **Chapter 3:**<br>**Learning by Figuring**<br>**It Out** | Help your students approach problems without first having a defined "correct" strategy. | Students<br>• use their **mistakes** to learn<br>• **surf** problem-solving strategies<br>• **predict and check** answers to a problem<br>• **work backward** from a given solution |
| **Chapter 4:**<br>**Learning by**<br>**Picturing It** | Help your students create images to explore problems and strengthen memory. | Students<br>• **visualize** a problem to "see" it<br>• **free sketch** to explore a problem<br>• create **concept maps** to find connections and relationships<br>• create **graphic organizers** to organize thinking |
| **Chapter 5:**<br>**Learning by Writing** | Help your students use informal, exploratory writing to help them think through mathematical concepts. | Students<br>• **paraphrase** a problem to help them understand it<br>• approach a problem by doing a **freewrite**<br>• do a freewrite of **30 words or less**<br>• use a **plus/minus chart** to record what they know and don't know |
| **Chapter 6:**<br>**Using Awareness**<br>**Strategies to Improve**<br>**Learning** | Help your students consciously improve their learning process. | Students<br>• **prioritize** aspects of a problem<br>• **summarize** a problem-solving experience<br>• **reflect** on a problem-solving experience<br>• use a **learning journal** to record key takeaways<br>• record their **next steps** |

**CHAPTER 1**

# Learning through Self-Assessment

*Help your students use objective tools to gain a realistic perspective of what they don't know yet and to consolidate what they do know.*

When we are on a journey, it is important to have a clear idea of our starting point, our destination, and our means of transport. There is no point in rushing out the door unless we know in which direction to travel. And, as we travel, it helps to know where we are, so that we can stay on track. What we need is a global positioning system (GPS).

**Too many students appear to be lacking a GPS for their learning journey.**

Too many students appear to be lacking a GPS for their learning journey. They may identify, or be told, what the learning goal is for a particular lesson or unit but be unsure where they are in relation to that goal. Students can better plan their route to mastery of a concept if they know what they know about the concept and what they do not know yet. This understanding must be objective and not based on inaccuracies. For students, the GPS they need for learning is effective self-assessment. Through ongoing self-assessment and feedback, students can gain a clear and accurate picture of their status and what their next steps should be.

Self-assessment (also referred to as calibration) is the act of using an objective tool to gain a realistic perspective of what we know and what we do not know (Brown, Roediger III, & McDaniel, 2014). By aligning our self-assessment with *objective* feedback, we avoid illusion of mastery.

To make self-assessment useful, we must guide students in using objective assessment tools. Using an objective tool enables students to avoid illusion of mastery because—like all of us—students are susceptible to inaccurate notions of what they know. When you provide students with objective tools, you will be helping them to get an accurate picture of where they stand.

To be effective, self-assessment tools must also provide timely feedback. Too often, when studying or practicing a strategy, students will move forward through countless problem sets without identifying if or where they went wrong. What this does is create two fallacies. First, without knowing if their responses are correct or incorrect, students will not have an accurate gauge of their understanding. They may think that they are strengthening their understanding of a concept when in fact they are not. What they are doing, instead, is reaffirming an inaccurate application of a strategy. Second, without checking their responses for accuracy, the student is forming a path not toward their learning goal but away from it.

Thus, to strengthen their ability to apply previous learning, students need assessment tools that provide feedback that is both objective and timely. Objective self-assessment tools can include self-quizzing, flashcards, study guides, question sets, self-monitoring, and so on.

Timing is key to successfully using self-assessment tools for learning. Taking a self-assessment directly following an initial lesson is not terribly helpful. The point should not be to measure short-term performance. Self-assessing is most useful when we take stock of our status after some time has passed, to assess what learning has "stuck."

Retrieval practice that is spaced and mixed provides students with the best opportunity to accurately self-assess. When time has passed and concepts are mixed, the learner is forced to think about the prompt, retrieve previous learning, reconsolidate knowledge, and then apply it to answer the question. This form of self-assessment will inform students of their progress in relation to their learning goals. It's about taking stock of their *retention* of understanding.

**Self-assessment tools are useful not only for *measuring* learning but for making that learning happen.**

Self-assessment tools are useful not only for *measuring* learning but for making that learning happen. While using a self-assessment tool may take more effort than rereading texts or notes, the greater effort exerted by the student will strengthen both the pathways to previous learning as well as the previous learning itself.

By self-assessing, students give themselves immediate feedback that can guide them in next steps. This may involve focusing on certain topics or changing the type of questions to practice. Self-assessment is an accurate guide for students to support them in becoming independent learners.

Some students may be able to create and answer questions that focus on the main ideas of a concept. Others will struggle to create suitable questions, so you may want to supply questions to prompt their thinking. After completing the self-assessment and identifying weaknesses, students can then address the gaps in learning with additional work.

Self-assessment tasks such as self-quizzing are powerful because students must work through desirable difficulties to recall previous learning and apply it to new contexts. To make this work, these tasks must be spaced over a period of time and must mix up the tasks being reviewed so that the same strategy isn't the focus of a single self-assessment experience.

### Using self-assessment to learn addition facts

I would work on the facts at school and at home. When I was asked questions in class, I couldn't remember my facts. I started to use the flashcards for addition and would put my work into two piles—what I knew and what I didn't know yet. If I kept answering a flashcard right, maybe like three or four times, I would put it into the "I know" pile; if I didn't know a flashcard right away, I would put it into the "don't know yet" pile. Then I would practice the "don't know yet" pile more than the "I know" pile. That helped me spend more time on the facts I didn't know.

I still practice both piles, but I practice the "don't know yet" pile more because that is what I need practice with.

**Building Retention**: This chapter introduces the following examples of Learning through Self-Assessment, which you can help your students use to build their retention capabilities:

- create their own **self-quizzes**
- use **flashcards** for mixed and spaced review
- **self-monitor**

## Students Create Their Own Self-Quizzes

**Suggested Prompts**

To inspire students as they create their own self-quizzes:
- *How could you decide which concepts to focus on?*
- *What type of questions could you ask yourself?*
- *What would you do if you responded to a question incorrectly?*

You may have to work with students to help them be comfortable responding to prompts like those above. For example, consider having students share their responses to these prompts so that classmates can hear other students' thinking. This is especially important as many students may not have encountered this approach to learning, or may not have a lot of experience with it.

What does "studying" look like? Most people would say it involves reading over notes and previous work. Self-quizzing is a different way to study that provides students with information that reading notes does not. Students may find the idea of self-quizzing unappealing because it requires more effort than reading over notes (Brown, Roediger III, & McDaniel, 2014). It is this additional effort, however, that strengthens both consolidation and retention.

During the process of self-quizzing, students must search their long-term memory for previously learned knowledge that relates to the specific questions being asked. What this does is strengthen that previous learning as students add context to the memory in the process of consolidation. It also strengthens the pathways to this learning.

For self-quizzing to be effective, students need an opportunity for immediate feedback. Feedback can take the form of the student checking their own work or having another student check their responses. This feedback will confirm to students if they were able to recall and apply previous learning. Feedback will help them gain an accurate picture of what they know and what they do not know. Students will then have a much better awareness of their progress and what next steps to take than they would have had if they only reread their study notes.

Depending on your students, you can provide an initial set of questions. If possible, though, have students design their own self-quizzes. Doing so will require them to think about which information to focus on. This creates an opportunity for them to sort through topics and identify the more prevalent ones. They can achieve this by either crafting their own questions or selecting questions from a text.

It will be important for you to support your students in developing their self-quizzing skills. Facilitate discussions on selecting content, question types, and the importance of validating responses. To be effective in self-quizzing, students will need time and support. Students may be tempted to choose questions they can answer easily. These would not provide them with the desirable difficulty that is a significant part in the learning process. Support these students in developing more complex questions that can give them a clearer snapshot of their learning.

The following three examples highlight how students can strengthen their ability to retain previous learning by self-quizzing and then getting feedback.

### PRIMARY EXAMPLE: STUDENT WORK

#### Self-quizzing counting by 2s

A student rolls two number cubes and then counts from that number by 2s. She writes down the number sequence and then uses a hundreds chart to check if her response is correct.

### ELEMENTARY EXAMPLE: STUDENT WORK

#### Self-quizzing two-digit multiplication

A student creates multiple two-digit by two-digit multiplication problems to solve. He solves the problems and then compares his work to the result found using a calculator.

### INTERMEDIATE EXAMPLE: STUDENT WORK

#### Self-quizzing key term definitions

A student records five key words from a math unit previously studied. He writes a definition for each from memory. The student then compares his written definitions with those in his notes to check for accuracy.

## Students Use Flashcards for Mixed and Spaced Review

Did you ever use flashcards to rote learn your multiplication facts? Many of us did. And yet, flashcards can be used for so much more. The trick is in *how* we use flashcards. By modelling a variety of ways to use flashcards in the classroom, you can emphasize the effectiveness of flashcards as a learning strategy. You can help students see that flashcards can help them consolidate and retain a wide variety of previous learning. Flashcards can focus on vocabulary, math facts, equations, graphs, patterns, and so on.

First, students can use flashcards to space review opportunities over a length of time. It's the kind of tool that is easy to return to time and again. This recurring retrieval of previous memories strengthens the pathways to knowledge.

Second, students can mix up the flashcards so that consecutive cards do not require the same strategy to solve the problem. When flashcards are unorganized, the review is mixed, thereby increasing cognitive demand for students. When seeing a new flashcard, students must pause to analyze the problem, consider which strategy is appropriate, and then apply their chosen strategy.

By setting aside flashcards they have answered correctly, students can focus on the questions and problems that they find more challenging. They should not, however, drop flashcards out of rotation too quickly. Answering a flashcard correctly once or twice does not guarantee that they will be able to recall this information in future. Students should answer a flashcard correctly at least three times, spaced over a period of time, to ensure that the memory isn't just a short-term memory (Karpicke, 2009).

Typically, teachers have been the ones who have created the content for flash-cards. This has its benefits, as the teacher can ensure all aspects of the concept are included. Having students create their own flashcards, however, has different benefits. During creation of the cards, students will have to access their previous learning for the finer details of the concept and find creative ways to apply it. Furthermore, these custom flashcards can be a classroom resource that is naturally differentiated to meet the learning needs of students.

Here are a few sample topics that can be addressed using flashcards:

- **Primary**—comparison, addition, subtraction, before/after, subitizing, patterns (increasing, decreasing, and repeating), and shapes
- **Elementary**—addition, subtraction, multiplication, division, place value, measurement (area, perimeter, volume), fractions, integers, and angles. Also relating decimals, fractions, and percentages.
- **Intermediate**—addition, subtraction, multiplication, division of fractions and integers, order of operations, solve for the unknown, and geometry

Flashcards can be used in math games such as Sort, Solve Snap, Solve War, Explain, Memory, Order, and Match. The following three examples illustrate how flashcards can be used effectively.

**DIGITAL SOLUTIONS**
You can create a digital flashcard set by making a slideshow, with each slide being a flashcard.

## PRIMARY EXAMPLE: STUDENT WORK

### Using flashcards to practice subitizing

A student used a set of flashcards with a ten frame on one side and a number on the flip side. After subitizing the number of dots on a flashcard showing a ten frame, the student checked the number on the back of the card.

## ELEMENTARY EXAMPLE: STUDENT WORK

### Using flashcards to practice multiplication facts

A student uses a set of flashcards that have a multiplication expression on the front and the product on the back. When the student responded correctly, he placed the flashcard in a "got it" pile. When he responded incorrectly, he placed the flashcard in a "not yet" pile. He continued to practice both piles, but the "not yet" pile would be practiced much more in comparison with the "got it" pile.

### Using flashcards to work with math terms and tasks

A student had two colors of flash cards. Green cards had a math term on the front and a definition on the back. The student went through the pile of cards one by one. For each math term, she recalled the definition and then looked at the back of the flashcard to check her response for accuracy.

Blue cards each had a task on the front of the card and a solution on the back. The student worked through the task and then compared her response to the solution on the back of the card.

## Students Self-Monitor

**Suggested Prompts**

To inspire students as they self-monitor:

- *Try talking out loud as you work through the problem: say what part you get and what parts you don't get.*
- *How do you know when you hit a stumbling block?*
- *How will you know if you are making meaning?*
- *How will you check to see if you're right?*

In too many classrooms, students ask their teachers, "Is this right?" or "How do I do this problem?" When the teacher answers such questions, it is the *teacher* who is thinking. By teaching students to self-monitor, you can empower *them* to do the thinking. You give students an ongoing, live feedback of their progress. They no longer must rely on you to tell them if they are right or not. It is the student monitoring their own thinking—a crucial aspect of independent learning.

Students self-monitoring for comprehension means that they are continually checking to see if their work makes sense as they work through a problem or task. It situates the student as the thinker. Self-monitoring students who run into roadblocks work to pinpoint if their confusion stems from the problem as a whole, from only part of the problem, or from an uncertainty about what approach to take. They try to identify the ideas, concepts, or themes that do not make sense.

Once self-monitoring students have identified areas of difficulty, they can then select an appropriate strategy to repair their gap in comprehension. They can turn to a variety of "fix-it" strategies, such as rereading the problem, making sure that all relevant information has been identified in a problem, checking to see if the strategy is being applied appropriately, and determining if the initial response even makes sense.

To help your students get the idea of self-monitoring, model the practice: as you demonstrate solving a problem, talk out loud and describe your thinking process. Purposely make mistakes and then talk yourself through your work to identify the errors. Explain that this is how you would like them to talk to themselves while they do problems. If your students need help getting comfortable with the practice, have them begin by trying to explain their thinking to you or a partner.

The following examples show students talking through their thinking or describing their strategies for self-monitoring as they work through a problem.

### Self-monitoring skip counting

I'm skip counting by 5s to 100. I'm starting by saying 10, 15, 20 … Now I just said a number that ends with an 8. I know that doesn't make sense because, when you count by 5s starting at 10, the last part of the number should always be 0 or 5.

ELEMENTARY EXAMPLE: STUDENT WORK

## Self-monitoring work on geometry problems

I'm working with translations, rotations, and reflections.... I'm not sure if my rotations are done right. Should I ask the teacher? No, I want to do it myself.... So, I'll start by putting a question mark beside anything that I'm not sure of and an exclamation mark beside things that I know for sure.... Okay, now I'll go back to the question marks and think about it a little more.... I can make some small changes.... Good, that's right now.

INTERMEDIATE EXAMPLE: A STUDENT EXPLAIN

## Self-monitoring work on fractions

I was solving some equations with fractions, and I wasn't sure if I was on the right track with my work. So I stopped and went back to the beginning of my work and rechecked all of my steps. I saw that everything was correct, so I kept going and solved the problem. I wanted to be sure that I was correct and not making any mistakes along the way.

## Learning through Self-Assessment Summary

Each of the strategies highlighted within Chapter 1—creating and taking self-quizzes, using flashcards for mixed and spaced review, and self-monitoring—are different ways of Learning through Self-Assessment. They can all play a significant role in helping students gain an accurate understanding of what they know well and what they do not know well. Further, all three strategies can help consolidate previous learnings. Instead of considering self-assessment as a measuring activity, the three strategies situate self-assessment as metacognitive activity that strengthens learning.

# 2

# Building a Network of Memory

*Help your students elaborate their knowledge of a concept by adding layers of meaning to a memory.*

When building with blocks, Kindergarteners don't take long to see intuitively which foundations are more stable than others. Consider two towers of blocks. The first tower consists of one block at the bottom with multiple blocks sitting one on top of the other. This tower can be quite tall; it does not, however, stand on its own very well and may collapse without encouragement. Consider a second tower of blocks that is the same height as the original tower. Its foundation, however, consists of many blocks; the tower gradually narrows toward the top. While this second tower is the same height as the first one, it is much more stable. It can stand on its own and will not fall unless pushed.

**Adding layers of meaning is like adding bricks to build a solid foundation. The more layers, the more stable will be the learning.**

The process of building block foundations can be an effective metaphor for the learning process. Elaboration, the process of adding layers of meaning to material (Brown, Roediger III & McDaniel, 2014), supports the student in strengthening their understanding of a concept. Adding layers of meaning (or elaborations) is like adding bricks to build a solid foundation. The more layers, the more stable will be the learning. When layers of meaning are supporting learning, students will have a deeper understanding of a concept and will have more links to retrieve memories about it.

Consider the example of addition in the primary grades. After introducing addition, you might help your students explore its connections to place value, representing numbers, part-part-whole, and the inverse of subtraction. You have given your students a deeper understanding of addition and many retrieval routes to the original memory of your teaching. What you have done is strengthen the memory, making it more comprehensive, and enabling your students to connect to the central concept through many avenues.

The more connections your students make, the more likely they are to remember relevant information. The more linkages amongst concepts, the stronger the learning becomes because it has been placed within a greater context in memory. Consider the following two diagrams. The first diagram consists of dots that are unconnected. Each of the dots can be thought of as a mathematical concept (such as addition, subtraction, place value, representing numbers, part-part-whole, etc.) that has been taught without making connections to the other concepts. With each dot being unconnected, there is no network to support understanding. Students cannot use any of the concepts to assist them in making meaning of any other concept. In this scenario, learning has to happen in isolation.

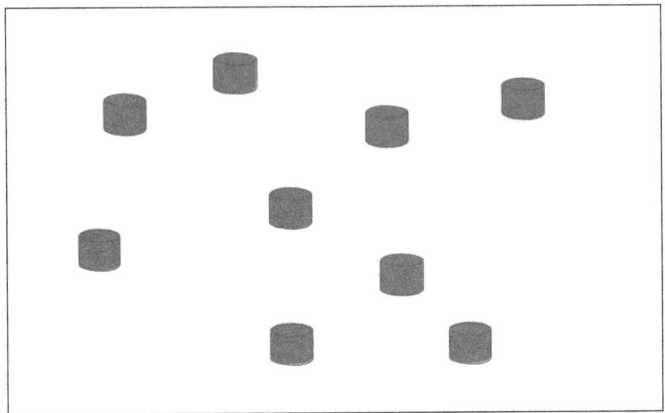

*When concepts are taught without connections to other concepts, they are harder to remember.*

**Connections give students the opportunity to rely on multiple concepts to make meaning for any other concept.**

The second diagram, below, consists of dots that are connected, forming a network. Again, we can consider each of the dots as a mathematical concept (such as addition, subtraction, place value, representing numbers, part-part-whole, etc.), but this time students are already aware or have been made aware of the connections amongst the concepts. These connections give students the opportunity to rely on multiple concepts to make meaning for any other concept. The network assists learning and supports students in digging deeper when they explore related concepts.

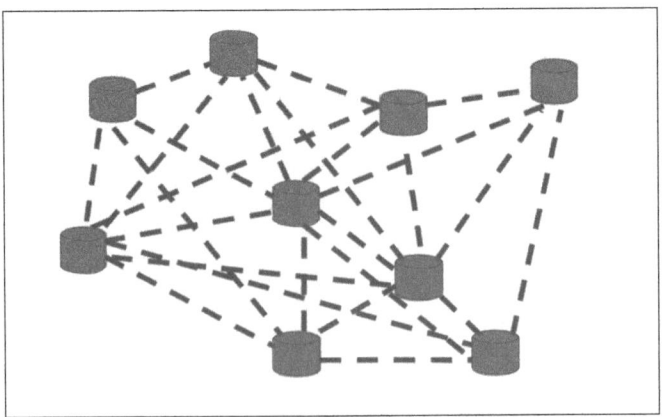

*When concepts are taught in connection to other concepts, they are easier to remember.*

Overall, elaboration can be summed up as adding more detail to memories of a concept and gaining the ability to situate the concept in a greater context.

So how can you build opportunities for elaboration in the classroom? You can introduce strategies such that students are generating examples of the concept, making connections amongst concepts, linking text with visuals, analyzing concepts for similarities and differences, asking why and how questions about a concept, and explaining the concept to themselves and others. These strategies, all of which we will explore, give students the chance to build additional layers of meaning to enrich their original understanding of a concept.

It is important for students to have feedback on their elaborations. You can encourage individual students to look through their work and check it against texts or modelled examples. You can ask students to share their work with a partner and provide each other with constructive feedback. Recognizing the level of accuracy of their elaborations will provide students with a realistic picture of their understanding.

To make the most of elaboration opportunities, space them out over a period of time so that students retrieve learning from earlier in the week or month. That way, the retrieval of memories about one concept can be interspersed with retrieval of other concepts. This mixing will entail desirable difficulty and increase the likelihood that students find connections to the other concepts they have been learning about. It is not as beneficial for students to elaborate on the same topic day after day.

---

**ELEMENTARY EXAMPLE: A STUDENT EXPLAINS**

### Comparing and contrasting area and volume

I kept getting area and volume confused. When I was given problems for area, I would try to solve them using volume. When I was given problems for volume, I would try to solve them using area. Then I was told to try to compare and contrast area and volume. I had to find things that they had in common and things that were different about them. I did this every few weeks, when we had reviews. It helped! I don't get area and volume mixed up anymore. When I read the problem, I think about what it is asking for and then decide on if this is area or volume. I get a lot more questions right now than before.

---

**Building Retention**: This chapter introduces the following examples of Building a Network of Memory, which you can help your students use to build their retention capabilities:

- create and answer **how and why** questions
- **explain** their thinking
- use **dual coding**
- **compare and contrast** concepts
- generate **concrete examples** of abstract ideas
- **make connections** to prior knowledge

# Students Create and Answer How and Why Questions

**Suggested Prompts**

To inspire students as they create and answer how and why questions:

- *What could you ask about …?*
- *What "why" question could you ask about …?*
- *Do you wonder what steps to take? Make a question asking about that.*

It is one thing for you to ask questions of students; it is quite another thing for students to create the how and why questions themselves.

Students can add layers of meaning to their understanding of a concept by asking and answering questions about it. More specifically, asking how and why questions provides opportunities for students to explore a concept deeply and to investigate the finer details within that concept (Weinstein, Sumeracki, & Caviglioli, 2019).

It is one thing for you to ask questions of students; it is quite another thing for students to create the how and why questions themselves. The latter requires much more cognitive effort. Another significant benefit to students comes from answering their own questions. By generating and answering questions about a concept, students will develop a more complex understanding of it, which in turn will improve the likelihood of remembering.

Consider the effort that would be required to come up with the following questions about perimeter:

- How do I find the perimeter?
- Why do I add the side lengths together?
- How will I know if my response is correct?

Students can create and answer questions orally, through text, or mentally. It is not important how students record their questions and responses (unless you want a record of this process). Students, though, must check to determine if their responses are correct so that they are building their learning on accurate information. They can do this by checking a text, reviewing their notes, or asking a classmate.

Encourage students to ask why and how questions about the concepts they are currently exploring but also of those they have explored in the weeks and months gone by. This forces students to examine their understanding of previously learned concepts. Through this consolidation, previous learning is strengthened as is the student's ability to recall and apply it.

You can help your students to become more comfortable with creating questions about concepts by making up questions as part of a classroom game, or by asking partners to help each other ask and answer questions about a concept or about their work. You may wish to record student questions as they are developed in the classroom so that students can use these as a springboard to generate additional questions.

The following examples demonstrate a few ways that asking questions about concepts can bolster student learning.

---

**PRIMARY EXAMPLE: A TEACHER EXPLAINS**

### Creating questions about patterns

The class was exploring patterns with two or three elements. Students seemed to know a little bit about patterns but wanted to know more. I asked them to think of a question they could ask about patterns. Afterwards, one student said, "I wondered how to recreate patterns with other items. So I made a question about how the pattern was repeating. When I figured this out, I made another pattern with blocks. I now can recreate patterns using other items."

---

### Asking questions about identifying expressions

I was doing input and output tables. I found it difficult to identify the expression for the pattern. So I kept asking myself why the output was always smaller than the input. This led me to think about division and subtraction, which helped me understand that output doesn't always have to be bigger than input. It depends on what operation you use.

### Asking questions about subtraction

I always liked integers when we were placing them on a number line. But, when we had to add and subtract integers, I didn't know what to do. My teacher said to ask myself questions about it. So I asked myself how does subtracting a negative number from a negative number have a difference bigger than the first negative number. Asking why and how made me understand that subtracting means "find the difference," which helped me see why the answer could be smaller than the first negative number.

## Students Explain Their Thinking

**Suggested Prompts**

To inspire students as they explain their thinking:

- *How would you explain the process you applied?*
- *What made your strategy effective?*
- *What strategies did you decide were ineffective? How could you tell?*

**DIGITAL SOLUTIONS**

Students could record their thinking digitally as they engage with a task or as a summary after the task is completed. They could record their explanations as text, a how-to video, or a podcast.

Self-explanation is the process of explaining to yourself or to others the steps you took to solve a problem. It is one thing to do something, it is another thing to remember and explain the process you used. To explain themselves, students are forced to consider carefully multiple aspects of a concept, as well as the sequential aspects of the problem-solving process.

When you engage students in self-explanation, your students have an opportunity to think about and sometimes share their thinking. Through this thinking, students are considering which aspects are important, why things work the way they do, and what this means to the overall picture. Self-explanation is an elaborative process by which students add details and layers of meaning to their memory of a concept and strengthen their ability to retrieve their learning about it.

You can create many opportunities for self-explanation that benefit both the explainer and the audience. You can ask students to share how they solved a problem to the class, to a partner, or in small groups. Students can write their explanations, but that is not necessary for the reinforcement of learning (this may depend on the student). Students could make a how-to video, a how-to podcast, or a how-to poster.

Explaining to others is one option, but the key is to get students doing this independently as a way to strengthen their learning. The following three examples show students engaging with self-explanation.

### Explaining how I measured

I love measurement. I had to order a few things from shortest to longest. When I was doing this, I explained the steps to my partner: I lined everything up beside the others. So I talked about which is the smallest of the things left and how that would go next in order.

### Explaining how I added decimals

We were adding decimals and some were into the tenths while others were into the thousandths. In my head, I was thinking about how to line the numbers up. I did it by place value, which helped me place all three numbers. It looked a bit weird because there were some digits without numbers under them. But, when I went through it step by step using place value, it was right.

### Explaining how I added fractions

I am getting better at adding fractions. I was a bit mixed up in the beginning, but then I talked to myself after I finished the first problem, explaining to myself how I did it. Then it got easier. I don't do this for every problem, but usually for the first two of the assignment. This helps me remember the procedure more.

## Students Use Dual Coding

**Suggested Prompts**

To inspire students as they use dual coding:
- *What picture comes to mind when you hear that term? Why?*
- *Draw a symbol that would remind you of …*
- *How would you describe the process displayed in the picture?*
- *What images could you draw to help you solve this problem?*
- *Think of some labels to add to your diagram. Use them to tell what is happening.*

Dual coding is an effective way to build multiple layers of meaning. Put simply, dual coding is the process of combining text information and visual information. Research supporting dual coding emphasizes that students learn better when they encounter concepts as a combination of words and visuals (Weinstein, Sumeracki, & Caviglioli, 2019). Words and visuals in combination are more powerful than either alone.

Dual coding enhances learning because words and visuals are processed through separate channels in the brain (Paivio, 1986). By processing the same information through two separate channels, the brain becomes more adept at retrieving this information later. Students have two avenues to follow when they need to remember information. Further, when examining both words and visuals, students tend to pay close attention to details in the two forms to determine the linkages between both. This process strengthens memory.

When dual coding, students should not be concerned with the quality of the visuals they create. The quality of the image is less important than the act of linking the concept, in word form, to a visual.

Some approaches to dual coding:

- thinking of terms or key concepts and creating an image of what it represents
- looking at visuals and trying to explain them in words
- identifying and labelling a visual

**English language learners will undoubtedly benefit from dual-coding strategies.**

You can encourage dual coding in the classroom through vehicles such as word cards. Have students create the word walls for a unit by writing a word on a card and drawing a picture and writing a definition on the flip side. When not on the word wall, these cards can be used for practice (as flashcards) but also for casual quizzing. They could also be great for a class, small group, or individual game of jeopardy. English language learners will undoubtedly benefit from such strategies.

**PRIMARY EXAMPLE: STUDENT WORK**

### Using dual coding to explore odd and even

A primary student creates pictures to pair with her definitions for the two terms *odd* and *even*.

**ELEMENTARY EXAMPLE: STUDENT WORK**

### Using dual coding to explore an algebraic expression

An elementary student uses words and images to better understand the expression 2x +1.

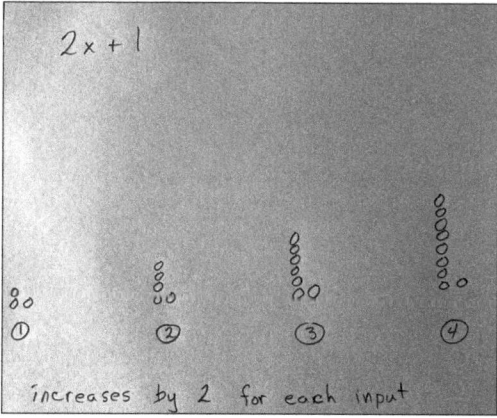

# Students Compare and Contrast Concepts

**Suggested Prompts**

To inspire students as they compare and contrast concepts:

- *What do these concepts have in common?*
- *What differences are there between these concepts?*
- *What organizer could you make to see how these concepts are the same and different?*
- *What criteria could you use to decide if these two concepts are more alike than they are different?*

Comparing and contrasting two concepts requires students to examine both concepts closely and to distinguish their similarities and differences. This process helps the student zero in on the foundational aspects of both concepts, thereby adding crucial layers of meaning to students' memories of the concepts. By adding these layers of meaning, students strengthen their learning because they are consolidating their memories each time they identify a foundational aspect.

There are multiple ways to encourage students to compare and contrast concepts.

- Provide a line master of a Venn diagram, which students can use to highlight the similarities and differences.
- Ask student groups to compile lists that highlight similarities and differences.
- For homework, students could write descriptions of two concepts that focus on the similarities and differences.
- In partners, students could take turns identifying similarities and differences.

All of these approaches will elaborate student understanding of concepts.

To incorporate spaced retrieval, encourage students to identify the similarities and differences of the same two concepts over a period of time. In addition, have students compare and contrast a variety of concepts as a means to mix the retrieval process and thereby increase the cognitive demand.

The following examples illustrate three different approaches to compare and contrast. The teachers had encouraged students to compare and contrast concepts that had not been addressed in the classroom for quite some time. Recalling information from earlier in the year supported students in strengthening the memory of a concept, thereby strengthening student ability to recall it.

---

**PRIMARY EXAMPLE: STUDENT WORK**

### Using a Venn diagram to compare and contrast skip counting by 2s and 5s

A student made a Venn diagram to highlight the similarities and differences between counting by 2s and 5s when starting at 0.

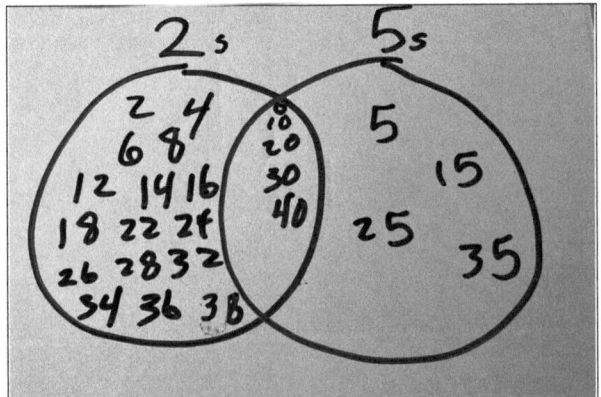

---

### Listing similarities and differences of area and perimeter

A student created this T-chart to identify the characteristics of area and perimeter. He then drew lines between similarities.

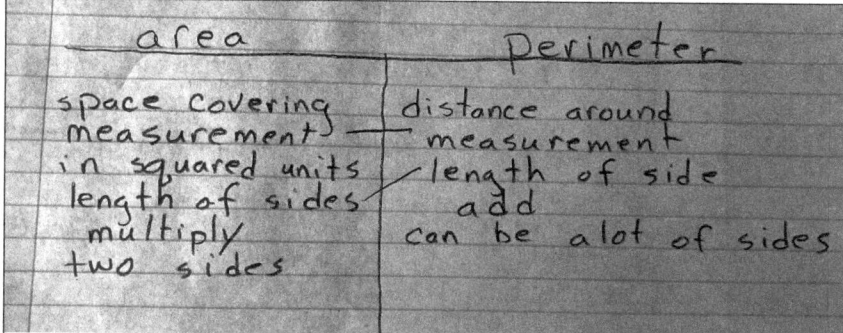

### Writing definitions of types of data to highlight similarities and differences

A student wrote these descriptions of continuous and discrete data to show how they are similar and different.

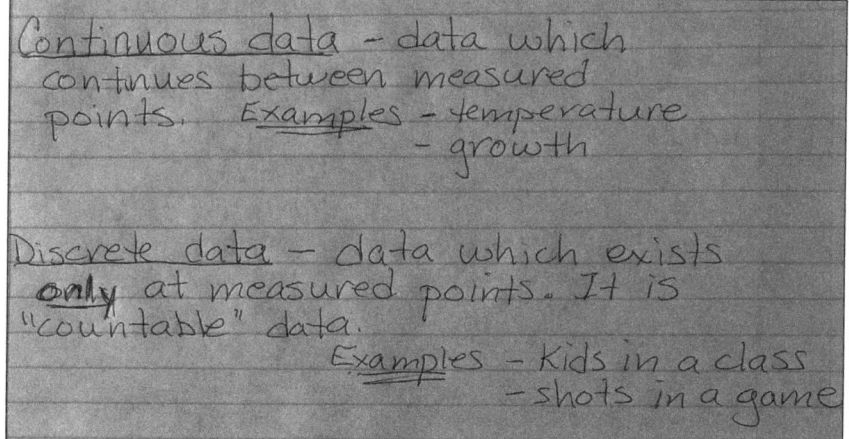

**Suggested Prompts**

To inspire students as they generate concrete examples of abstract ideas:

- *What comes to mind when you think of that concept?*
- *Have you seen this concept at work in the world?*
- *Perform a skit in which the main character is obsessed with a real-world example of …*
- *For homework, take a picture of … that you see every day.*

## Students Generate Concrete Examples of Abstract Ideas

When sitting in math class, students may feel that math is quite abstract. And when concepts seem abstract, it can be difficult to make meaning. To address this situation, a teacher can give concrete examples of the concept and—even better—encourage students to come up with their own concrete examples. Humans are better able to recall concrete information than abstract ideas (Weinstein, Sumeracki, & Caviglioli, 2019). By connecting the abstract concept to a concrete example, students can add meaning to their memory of a concept, add additional details to their understanding of the concept, and have greater success in recalling it later.

When students are trying to come up with concrete examples for abstract ideas, encourage them to draw examples from their own lives and experiences. These will be easier to remember. For example, the concept of area can be difficult for some elementary students to grasp. To support them with this abstract notion, ask them to think of areas in their own lives, such as their bedroom floor, the surface of a kitchen table, or the viewing area on a TV screen. In addition, you can encourage students to think of actual real activities that involve area in the real world. For instance, area needs to be measured when buying carpet, purchasing seed for a yard, painting surfaces, and designing outdoor spaces. Each of these examples brings meaning to the abstract idea conveyed by an equation for area.

Thinking of concrete examples for abstract ideas can be encouraged during any part of the learning journey. You can make it part of classroom discussions and assignments. You can incorporate it as part of homework in the evening, review before a quiz, and study at the end of term. Generating concrete examples at any stage will strengthen students' memories and help them to access it more easily later.

**DIGITAL SOLUTIONS**
Students could post in a class web page photographs or scans of drawings that represent abstract mathematical ideas.

---

**PRIMARY EXAMPLE: STUDENT WORK**

### Picturing a seesaw to represent equality

A teacher asked students to think of an image to remind them of the meaning of the equality symbol (=). One student thought of a seesaw. If one side had more on it than the other side, the seesaw wouldn't be balanced. It was only when both sides had the same amount that the seesaw was balanced and equal.

---

**ELEMENTARY EXAMPLE: STUDENT WORK**

### Remembering the penny to reinforce rounding

To better remember how rounding works, a student thought of how the penny was no longer used in Canada and that cents had to be rounded to the nearest 0, 5, or 10. This concrete example helped to provide context and to serve as a reminder of how rounding works, especially when using decimals.

---

**INTERMEDIATE EXAMPLE: STUDENT WORK**

### Picturing a thermometer to reinforce negative integers

A student was having difficulty understanding and ordering integers, especially negative integers. To help herself remember the order of negative integers, she thought of them in terms of temperature below zero. She remembered that the colder it was, the lower the negative temperature would be. This helped her make meaning of negative integers when ordering.

# Students Make Connections to Prior Knowledge

**Suggested Prompts**

To inspire students as they make connections to prior knowledge:
- *What concept does this remind you of?*
- *How does this concept relate to other concepts?*
- *What connection can you make to what you already know?*

Connections provide the linkages that support student learning. When engaging with a mathematical concept, each student brings with them their collective prior experiences. It is through the purposeful recall of relevant prior knowledge that students make meaning, whether it be before, during, or after they engage with a concept. Students can draw relevant prior knowledge from their personal experiences, their understanding of the world, or earlier encounters with other mathematical concepts. It is by activating prior knowledge that students make connections and deepen their understanding of a concept beyond the superficial level.

Connecting to prior knowledge can have a huge impact on the inferences students make while working through a problem. It enables students to use background knowledge to support and internalize their understanding of a new concept. By making a connection, students can engage with new information without starting from scratch and they are more likely to remember the concept in the future.

You can encourage students to make connections to prior knowledge at all points of their learning, be it during the introduction of a concept, their exploration of it, or their attempt to recall it later. For the purpose of spacing and mixing review, creating and recalling these connections can be spread out over the term to facilitate both spaced and mixed review.

Let's consider how you might encourage the making of connections during a unit on integers. Perhaps you could help students recollect previous materials that students would be familiar with, such as the number line, double-sided counters, and so on. You could help them make connections to concepts such as addition and subtraction, temperature, money, and depth. To encourage deeper thinking, have different groups of students describe how integers relate to these topics, and then have students do a think-pair-share activity, a class show-and-share, or a strategy harvest.

---

**PRIMARY EXAMPLE: STUDENT WORK**

### Accessing understanding of part-part-whole to distinguish between addition and subtraction

A student kept mixing up addition and subtraction. The teacher suggested bringing to mind the part-part-whole diagram, which the class had learned about the previous week. The student figured out that if the whole was missing you had to add, and if a part was missing you had to subtract. The memory of the diagram later helped the student remember the difference between addition and subtraction.

---

## Using knowledge of race finishing times to remember how to order decimal numbers

Ordering decimal numbers can be challenging. After thinking about the numbers to the right of the decimal, a student remembered seeing times for races that were similar. And she knew that if the digits to the left of the decimal were the same, then you had to compare the digits just to the right of the decimal first. This helped the student order decimal numbers because she thought of it in terms of measured time in races.

## Remembering how a circle graph works by thinking of it as a pizza

To remember how circle graphs work, an intermediate student thought of them in terms of a pizza. The size of the slice indicates how much of the whole pizza it represented. This idea helped the student remember that the bigger the "slice" of a circle graph, the greater amount it represents in relation to the whole.

## Building a Network of Memory Summary

Each of the strategies in Chapter 2—creating and answering questions, explaining thinking, using dual coding, comparing and contrasting, generating concrete examples of abstract ideas, and making connections to prior knowledge—support students in Building a Network of Memory. While each of the examples approaches elaboration from a different angle, all of them assist in adding connections that strengthen students' understanding of a concept and increase pathways to their memory of it.

# CHAPTER 3

# Learning by Figuring It Out

*Help your students approach problems without first having a defined "correct" strategy.*

When we try to learn a new skill or concept, we have two choices: We can follow someone else's example, or we can try to figure it out ourselves. Let's consider the first option. If we watch an expert fix a portable heater, for example, we will see the process they apply and we will see them solve the problem. However, we may miss crucial details and we may not see the reasoning going on in the expert's brain.

Let's consider the second option—taking the heater apart ourselves and figuring out what the problem is. We may struggle though parts of the process to determine the appropriate strategy and next step: Should we take the back off first or the bottom? We may encounter mistakes as we work by trial and error: We might try using a flat-head screwdriver. When that doesn't work, we might look more closely and realize that we need a Phillips.

> It is through mistakes that we learn about a problem and process more deeply. We must pause and take stock of why something did or did not work, and what our next step should be.

Mistakes, however, are valuable for learning. It is through mistakes that we learn about a problem and process more deeply. We must pause and take stock of why something did or did not work, and what our next step should be. Struggling to take a portable heater apart and fix it means that we experience the multiple thinking points that we could miss while watching an expert fix it.

The above analogy applies very well to math. Too often, students observe a teacher work through a problem; they *see* the steps, but they do not see or experience the thinking that supports the teacher's actions. When students work through the problem themselves, they benefit by working through any misconceptions that they may have. Doing so is comparable to experiencing a desirable difficulty, or productive struggle. When you have students solve a problem without first showing them the correct approach, they will have to think about the nature of the problem and what strategies might work and why.

It is far better to attempt to solve a problem than to memorize a given solution.

Generation (or "figuring it out," as I will refer to it from here on) is the process whereby your brain attempts to solve a question or task before you understand what the solution would be (Brown, Roediger III, & McDaniel, 2014). Then, once you determine the solution or if corrective feedback is provided, your brain will adapt to the new information due to the effort created while seeking the solution.

Making mistakes and then recognizing when you need to correct those mistakes builds the bridges to stronger learning (Brown, Roediger III & McDaniel, 2014). It is far better to attempt to solve a problem than to memorize a given solution. When given a solution first, we tend to focus on following the steps at the expense of grasping the understanding that supports those steps.

Students who follow a prescribed formula to get an answer promote their short-term performance but do not support their learning process over the long run. Learning for the long term happens when students search their previous learnings to find connections and possible approaches, devise a plan of action, and then try it out. By trying various approaches, students zero in on the appropriate solution by eliminating non-efficient strategies.

The effectiveness of the "figuring it out" learning strategy will be strengthened if you alternate between providing a solution and not doing so, and if you space out these challenges. By mixing the approach and spacing out opportunities, you will support students in retaining and being able to recall their learning.

---

**INTERMEDIATE STUDENT: A STUDENT EXPLAINS**

### Figuring out how to subtract fractions

I never liked fractions at the beginning. I can remember when we learned to add fractions: the teacher showed us samples and then we used the same way to solve new ones. I thought I knew how to do it, but I guess I didn't. Then, when we were given fractions to subtract, we had to figure it out on our own. I didn't want to at first, but then I tried a few things. Didn't work. So I tried to solve the problem by making pictures of the fractions first, and then I subtracted. I was right!! It has been a while since we did subtracting fractions, but, when I was given two fractions to subtract yesterday, I was able to do it because I remembered how to use pictures to solve it. I also can add fractions better now than before because I use the strategy I found for subtraction.

---

**Building Retention**: This chapter introduces the following examples of Learning by Figuring It Out, which you can help your students use to build their retention capabilities:

- use their **mistakes** to learn
- **surf** problem-solving strategies
- **predict and check** answers to a problem
- **work backward** from a given solution

## Students Use Their Mistakes to Learn

Making mistakes was once considered to be undesirable by educators and students. Research has since demonstrated that making mistakes is an active part of learning and that we learn from mistakes (Boaler, 2016; Dweck, 2006). In fact,

**Suggested Prompts**

To inspire students to use their mistakes to learn:

- *That's a good idea. Go ahead and try that—we'll see if it works.*
- *What mistake did you make that helped you understand the concept better?*
- *You took a risk. And you made a mistake. Great! So you know that doesn't work. What could you try next?*

when teachers convince students that they can learn from their mistakes, students see the power in working through a problem and overcoming stumbling blocks.

When they have a growth mindset, students are more likely to take risks in their learning and then view their mistakes as exercises for their brain. Students learn valuable information when they make mistakes. They identify misconceptions they can address in their next attempt. They can narrow their focus on an appropriate strategy and apply it. Their learning of the concept is deepened by looking at it from different perspectives (changing a misconception to an accurate perception). Students' ability to retain their learning is thereby increased.

In the classroom, we typically do not plan a lesson around a mistake ahead of time. However, planning instruction with misconceptions in mind can be helpful. If you prepare your students to view mistakes as part of the learning process, your job is half done. But you also must be prepared to catch those "learning moments" and to use them to help individual students—or the class as a whole—to process those mistakes as the opportunities that they can be.

---

**PRIMARY EXAMPLE: STUDENT WORK**

### Learning from mistakes when adding two-digit numbers

A student was adding two-digit numbers and didn't want to know what the answer was. He wanted to figure it out himself. The student was forgetting to regroup when adding. To address this misconception, he started to draw visuals of base-ten blocks and then regroup 10 ones into 1 ten. Though it took a few times, each time he made a mistake he pointed it out to himself and said he knew not to do that again. Over time, the student could solve all the problems without making mistakes.

---

**ELEMENTARY EXAMPLE: STUDENT WORK**

### Learning from mistakes with expanded and standard form

A student was having difficulty with expanded form and standard form. He was mixing it up. He would do some work, ask for it to be checked, and then discuss his mistakes with his teacher. From there, the student started to understand expanded and standard form and learned to fix mistakes he made by noticing when things didn't make sense.

---

**INTERMEDIATE EXAMPLE: STUDENT WORK**

### Learning from mistakes while dividing fractions

Dividing fractions can be challenging for students. One student tried solving the problems on her own and then checked her answers by looking at the solution sheet. She identified problems she got wrong and then looked through her steps again to see what kind of error she was making. She soon figured out that she was not inversing the fraction before multiplying it. So, for $\frac{1}{2}$ divided by $\frac{1}{4}$, the student multiplied $\frac{1}{2}$ by $\frac{1}{4}$ to find $\frac{1}{8}$ instead of multiplying $\frac{1}{2}$ by $\frac{4}{1}$ (the inverse of $\frac{1}{4}$) to find 2. Afterward, the student was able to check for this kind of mistake when she was dividing fractions.

# Students Surf Problem-Solving Strategies

**Suggested Prompts**

To inspire students as they surf problem-solving strategies:

- *Which strategy will you try first?*
- *How will you use your first attempt to help you with your second?*
- *How will you know if this strategy is the best one to use?*

The surfing strategy for solving problems means giving students the freedom to try out a variety of approaches to a problem to see which one works. This may seem like a simplistic approach, but it has many strengths.

First, surfing problem-solving strategies allows the student to enter the problem even when they are unsure which approach is the right one. It provides a risk-free entrance into the problem.

Second, surfing problem-solving strategies supports students in strengthening their reasoning when engaged with a concept. In virtually any attempt, students receive feedback that can help them identify if their guess is suitable in solving this type of problem. This feedback will assist them in making a more informed attempt the next time. Through successive attempts, they will work their way toward an appropriate strategy and then to a solution. The premise is that students will use errors to inform future attempts.

Third, surfing problem-solving strategies assists students in adding more detail to their previous learning. Each time they try out a different problem-solving strategy, the student will return to the learning and consolidate their new experience with previous learning. This will strengthen the memory and the pathways to this memory.

For example, students may surf problem-solving strategies to find the pattern rule (or expression) when working with input/output tables. They will work through the four operations to see which one is suitable, and then refine their decision-making.

The following examples of surfing problem-solving strategies emphasize its effectiveness as a strategy to support learning in the classroom. In each example, the teacher encouraged students to surf various problem-solving strategies.

---

**PRIMARY EXAMPLE: STUDENT WORK**

### Surfing strategies to extend a pattern

Students were asked to extend a pattern consisting of four elements. One student was unsure. The teacher suggested that he surf strategies to find one that works. The student first guessed that the pattern was increasing. He recorded the next three elements in the pattern, but, when reviewing his work, he realized that the pattern wasn't increasing. Instead, it was repeating. From this, he identified the core of the pattern. The student circled the elements that were repeating within the pattern. He then extended the pattern by recording the four elements that continued the pattern. This second attempt was correct.

---

**ELEMENTARY EXAMPLE: STUDENT WORK**

### Surfing strategies to calculate area

A student was working on a word problem that asked how much floor tile was needed to tile a kitchen. The teacher had asked students to surf to find a strategy to use. Initially, the student guessed that she would have to add the dimensions to find the amount of tile. When she tried that out, she realized that this was incorrect. From there, the student realized that if adding the length of the sides to find the area wasn't correct, then she should try multiplying them instead. The first guess served to inform her second attempt.

---

**INTERMEDIATE EXAMPLE: STUDENT WORK**

### Surfing strategies to solve a word problem involving fractions

A student was solving a word problem that involved fractions. He was unsure of what to do, so the teacher suggested surfing problem-solving strategies. He guessed addition was the necessary operation to solve the problem. The sum, however, didn't make sense. The student thought about this and realized that subtraction would work better. He solved the problem after the first attempt failed, allowing him to narrow his selection of strategy.

---

## Students Predict and Check Answers to a Problem

**Suggested Prompts**

To inspire students as they predict and check answers to a problem:
- *Without doing any calculations, what do you predict might be a reasonable answer to this question? How do you know?*
- *What range do you expect the answer will fall within? How do you know?*
- *Did your response coincide with your prediction? Why are they different?*
- *What strategies could you use to predict the solution? Maybe start by reading the problem carefully for clues.*

Predicting entails making inferences and then projecting these forward to identify possible outcomes. In mathematics, we may not always use the term *predict*, but we often use other terms to mean the same thing. When asking students to estimate, for example, we're asking them to predict. When asking students to state the next element in a pattern or the next data point on a graph, we are asking them to make a prediction.

Unlike a guess, a prediction should be founded on evidence. The skill of making predictions may involve both articulating and rationalizing a prediction for assessment and refining a prediction, if necessary. When making a prediction, students analyze the information offered in the initial question. When refining a prediction, students are analyzing the results of their calculations or observations. Both activities serve to strengthen memory and build retention ability.

You can model predicting for your students whenever you work through problems with them. Before demonstrating the "correct" approach, identify evidence—the given information—and settle on a reasonable prediction. For subsequent problems, work through the prediction process with students, and then work through the problem together to check the prediction. After that, they should be able to predict and check problems themselves.

The following examples provide insight on how students can apply predictions to strengthen learning.

---

**PRIMARY EXAMPLE: STUDENT WORK**

### Predicting whether various 3-D solids roll

Students were provided 3-D solids to sort using one attribute. One student chose to sort based on if the solid would roll on a slope. For each solid, she made a prediction and then tested her prediction by placing the solid on a slope. Each prediction helped her envision what would happen before the actual experiment. The combination of prediction and confirmation strengthened her learning about the attributes of various shapes.

---

> ### ELEMENTARY EXAMPLE: STUDENT WORK
>
> ## Predicting based on theoretical probability
>
> A student was working on probability and had to make a prediction of what color tile he would pick out of a mystery bag. Equal numbers of red, green, and yellow tiles had been placed in the bag—6 of each. After 1 red, 5 green, and 3 yellow tiles were drawn, the student made a prediction. He said red because he knew there were more red tiles left in the bag than the other two colors added together.

> ### INTERMEDIATE EXAMPLE: STUDENT WORK
>
> ## Predicting the alignment of theoretical and experimental probability
>
> An intermediate student was working with the phrases theoretical probability and experimental probability. The student used a coin toss as an example. She knew that the theoretical probability was 50-50. She wanted to see if the theoretical probability and the experimental probability would get closer as she increased the number of trials. The student predicted that it would. Initially, into the trials, the student had results that showed significant differences between theoretical and experimental probability. But, after 20 trials, the two probabilities began to coincide.

## Students Work Backward from a Given Solution

**Suggested Prompts**

To inspire students as they work backward from a given solution:
- *What strategy would make sense, given that answer?*
- *How do you know if your strategy will lead to the solution?*
- *Is there more than one strategy you could try?*

To solve a problem, we usually start with facts, find the right strategy, and move toward the solution. But there is something to be said for doing it the other way around: starting at the solution and working our way backward to figure out the strategy. By providing your students the opportunity to start from the finish, you give them the opportunity to consider problems from a different perspective.

This different perspective—seeking a strategy after knowing the answer—adds a desirable difficulty because we do not typically approach problems from this direction. When starting with the solution, students can think about which strategies would have been effective in helping them reach this correct response and which strategies would have led them astray. At the same time, starting with a solution provides parameters for the students' thinking because it confines possible strategies to those most likely to result in the given answer.

Working backward goes beyond the traditional methods for getting information into heads. It is another approach you can use to add layers of meaning to students' memories of a concept. Further, it gives students a better appreciation of what they know and what they don't yet know.

You can incorporate the working-backward approach to problems particularly effectively during mixed review, when students must work harder to recall strategies that might work. Additionally, you can have students work on review together by trading question and answer sets and by challenging each other to find the correct strategy.

## Working backward to solve addition and subtraction problems

When working with addition and subtraction of three-digit numbers, a student first looked at answers provided by the teacher. Then he chose and applied a strategy that he thought would lead him to those answers. The answers provided by the teacher were like a map, helping the student determine if his chosen strategy was leading him to a solution or an error.

## Working backward to solve multiplication and division problems

When working with multiplication and division of decimals, a student used a calculator to find the answer. Then she applied a strategy she thought would lead her to this answer. The answer provided by the calculator was like a guidepost keeping her on track for her destination.

## Thinking of a solution, and then working backward to find the right strategy

An intermediate student, when provided a problem, would think of possible solutions that made sense. He then selected a strategy he thought would work. This backward-thinking approach helped him narrow down the list of strategies to try.

## Learning by Figuring It Out Summary

Each of the strategies in Chapter 3—using their mistakes to learn, surfing problem-solving strategies, predicting and checking answers to problems, and working backward from a given solution—are different ways of Learning by Figuring It Out. The more students work with a concept in different ways, the more likely that they will retain their understanding of it.

# 4

# Learning by Picturing It

*Help your students create images to explore problems and strengthen memory.*

**Our brains visualize the unknown to understand it better. It's how we make meaning.**

Think about when we read or hear about something unfamiliar. One of the first things we do is start to imagine—to picture it in our minds. In some instances, we think of this in relation to our own experiences and try to shape the image based on this. We revise and add to this image as more details are shared with us. Our brains visualize the unknown to understand it better. It's how we make meaning.

Visualization involves thinking in pictures. It is the act of creating images in our head based on what we hear or read or think about. Visualization is a learning strategy that supports understanding. Applied to mathematics, visualization is the process of representing concepts as mental images, thereby allowing the student to remember and manipulate concepts when making meaning (Small, 2013).

By encouraging visualization in connection with a mathematical concept, we inspire students to add layers of meaning to their learning, thereby strengthening the memory and enriching their whole learning experience. By visualizing concepts, students make connections with other concepts or personal experiences. The more practice a student has visualizing, the more automatic it becomes as they encounter new information, in the form of vocabulary or strategies.

Typically, people think of geometry, measurement, and graphs as good opportunities for visualization in math. Visualization can play a powerful role, however, in developing the ability to retain virtually any math concept. For example, when first learning numbers, many students create mental images as they learn the number symbols.

**Suggested Prompts**

To inspire students to picture concepts:
- *Where have you seen this concept in the real world?*
- *How can you represent this learning visually?*
- *Draw a picture of what you're thinking.*

**When you offer students opportunities to make meaning as they visualize concepts that are abstract, complex, or novel, you give them a rich learning experience that strengthens their understanding.**

When students apply visualization to their learning, they are working with different areas of the brain than if they only used numbers and words (Park & Brannon, 2013). Working with different areas of the brain strengthens student mathematical learning (Boaler, 2016). By using visualization, students will create a more detailed memory of the learning and will build more pathways to that memory.

The creation of concrete images can also greatly aid in the learning process. At its simplest, students can draw images to build their understanding. Representing their learning in concrete images provides students an opportunity to ponder, add detail, label, and identify the finer details of a concept that they may not notice when only writing or using numbers. Such drawings can support the student in both developing their learning and demonstrating their understanding.

Graphic organizers provide a different kind of visual support—they help students organize their thoughts. An abundance of graphic organizers exists to scaffold learning. They can support students in making meaning, identifying similarities and differences between concepts, highlighting key aspects of concepts, demonstrating the flow of ideas throughout a process, and more.

Too often, however, graphic organizers are thought of as places to store knowledge. Instead, we want students to realize that graphic organizers are tools to aid them in working through a concept and strengthening their understanding for the purpose of recall and applying knowledge to novel situations. When students understand that completing a graphic organizer is not the goal but a means to support their understanding, they will get a lot more out of using them.

A multitude of opportunities exists for students to visualize, either mentally or written. Similarly, there are an abundance of visual supports, in the form of graphic organizers. By mixing and spacing the use of these tools and strategies in your classroom, they can help you strengthen student understanding, in terms of both memory and retrieval.

---

**ELEMENTARY EXAMPLE: A STUDENT EXPLAINS**

### Using visualization to see a pattern

I was working on identifying the expression for a table of values. In the table, I had all the inputs and most of the outputs. There were six inputs but only four outputs—I had to calculate the missing outputs once I found the expression for the table. I was having a hard time—I couldn't get the expression. So I started to think about each ordered pair in my head. I was drawing a picture of the output for each input. After a few pictures, I could tell that the output was 3 times the input and then 2 less. So 3b − 2 was my expression. I couldn't figure it out just looking at the numbers, so I had to think of pictures in my mind. That helped.

---

**Building Retention**: This chapter introduces the following examples of Learning by Picturing It, which you can help your students use to build their retention capabilities:

- **visualize** a problem to "see" it
- **free sketch** to explore a problem
- create **concept maps** to find connections and relationships
- create **graphic organizers** to organize thinking

## Students Visualize a Problem to "See" It

**Suggested Prompts**

To inspire students as they visualize a problem to "see" it better:
- *What do you see when you think about this concept?*
- *What do you picture when you think about the problem?*

Remember this old saying: "A picture is worth a thousand words"? In math, seeing or visualizing a concept can help the brain process information to deepen understanding. Visuals can add layers of detail and context to mathematical understanding that might take a thousand words to otherwise explain. As students learn more about a concept, they adjust their mental images to accommodate the new information they are processing. This accommodation of new information strengthens the understanding of a concept and will support students in their ability to recall it later. A significant aspect of visualization is the ability of the student to articulate how the images they create in their mind supports their understanding.

Visualization has received significant attention in the world of mathematics (Small, 2008). Frequently, it has been associated with students drawing pictures to help them work through a problem. Students can accomplish the same benefit from images simply by visualizing. When problem solving, for example, visualization can help in framing the problem, working through the problem, and communicating the solution. In this way, visualization is about using mental images and representations to create, adjust, retrieve, and communicate understanding.

In mathematics, there are three purposes for visualizing.

- First, visualization supports the problem solver in entering and framing the problem. It supports understanding by enabling the problem solver to clarify any conceptions they have of the problem before selecting and applying an approach.
- Second, the problem solver uses visualization to model the problem. It may be unrealistic to use physical representations of the problem, so the problem solver relies on a mental image to model the problem being explored.
- Third, using visualization as a support allows for the problem solver to question the outcome. By doing this, the problem solver can work through different scenarios to identify patterns or oddities.

An easy way to get your students started with visualizing is to present a problem to the class, and then have students close their eyes to create mental images to better "see" the problem. Another approach is to encourage students to draw their thinking as they work through a problem.

### PRIMARY EXAMPLE: STUDENT WORK

#### Visualizing a hundreds chart to help with skip counting

A student was having difficulty remembering how to count by 2 starting at 0. So the student visualized a hundreds chart and mentally skipped over every second number. The student said that picturing the hundreds chart in his mind helped him keep track of the numbers.

ELEMENTARY EXAMPLE: STUDENT WORK

## Visualizing base-ten blocks to help with multiplication

When working with multiplication of two-digit by one-digit numbers, a student attempted to use an array to help with understanding. For this particular student, though, the use of an array did not aid in understanding the problem or in reaching the solution. What eventually helped him understand the process of multiplying was visualizing base-ten blocks.

INTERMEDIATE EXAMPLE: STUDENT WORK

## Visualizing fraction strips to subtract fractions

Subtraction of fractions with unlike denominators can be challenging. An intermediate student shared how she would picture fraction strips in her mind as she was subtracting. She pictured removing just the portion of the fraction needed. This enabled her to visualize the difference. For example, in $\frac{9}{10} - \frac{2}{5}$, the student visualized $\frac{9}{10}$ and then removed $\frac{2}{5}$ from that visualization. This took two steps: first she removed $\frac{2}{10}$ and then another $\frac{2}{10}$ (for a total of $\frac{4}{10}$). This left the difference of $\frac{5}{10}$ remaining. Maintaining a visual as she worked helped her understand the process of subtraction.

## Students Free Sketch to Explore a Problem

**Suggested Prompts**

To inspire students as they free sketch to explore a problem:
- *What picture do you see as you try to understand this problem?*
- *Make a sketch to help a classmate understand ...*
- *Use pictures to show how you solved the problem.*

**DIGITAL SOLUTIONS**

Some students might embrace free sketching more willingly if they can use graphics software to make their sketches.

When considering visual supports as a learning strategy, it is important to think of how students make meaning through images. One of the best strategies is the simplest: provide as little restriction as possible and encourage students to free sketch their interpretation of a concept or problem-solving experience.

Think of a free sketch exercise as a drawing-to-learn exercise. Students create an informal image, not meant to be neat or to demonstrate one's comprehensive knowledge about a concept. Instead, students create images to highlight a connection between two concepts, represent the layers of detail within a concept, show important aspects of a problem-solving experience, or share stumbling blocks on the way to understanding. While creating these informal sketches, students will retrieve previous learning and consolidate it with what they have learned since or with what they are currently exploring.

Free sketching can strengthen learning particularly effectively when it is spaced and mixed. By spacing opportunities to sketch their learning, students have an opportunity to revisit concepts over a period of time, thereby strengthening their ability to recall the learning. Mixing the opportunities to sketch will provide students with an extra step in their thinking because they have to pause to consider which sketch and which details will best represent the concept.

## Free sketching to help think through subtraction

A primary student quickly drew a picture to represent the subtraction situation to help her solve a subtraction question.

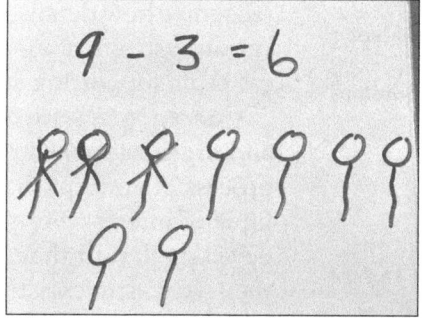

## Free sketching to think through equivalency

An elementary student drew a quick picture to think through why the improper fraction and mixed number are equivalent.

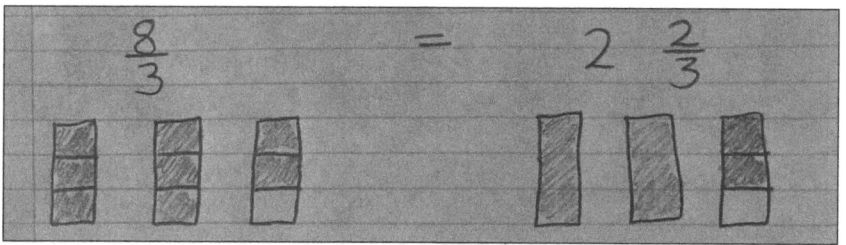

## Free sketching to illustration a geometric relationship

An intermediate student drew a quick sketch to show the relationship between radius and diameter.

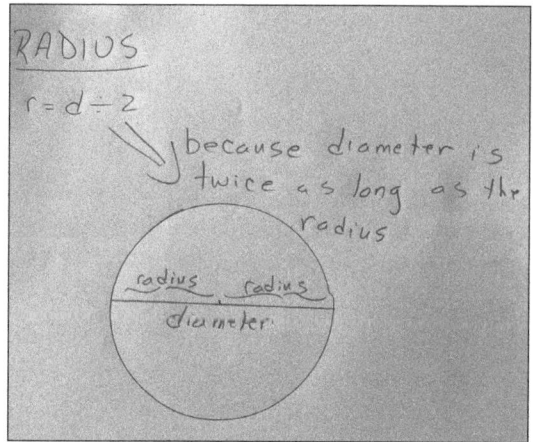

# Students Create Concept Maps to Find Connections and Relationships

**Suggested Prompts**

To inspire students as they create concept maps to find connections and relationships:
- *How do the parts of this concept relate to one another?*
- *How could you show relationships in this concept web?*

**DIGITAL SOLUTIONS**
Word processing software can be used to create concept maps.

The concept map is an excellent visual support for learning. By creating a concept map, students develop, strengthen, and demonstrate their understanding of a concept. Students must consider the foundational aspects of the concept and recognize how details relate to one another. Concept maps can be focused on the various aspects of an individual concept or on the relationships among linked concepts supporting an overarching topic.

Concept maps are diagrams that require the learner to consider the relational and organizational structure of a concept or topic (Novak, 2013). To begin the process of concept mapping, students brainstorm all that they know about the topic of interest. From there, students organize their gathered information and develop a layout that illustrates the relationship between various aspects. From the layout, students then describe the relationships with key words or phrases to highlight the interconnectedness. You may need to model ways to highlight the relationship of concepts within the concept map, for example, using colors, arrows, and shapes of bubbles.

Overall, concept maps highlight significant relationships amongst concepts within an organized structure (Karpicke & Blunt, 2011). They help students strengthen their learning through an active learning task that adds layers to memory. The act of creating the concept map helps students recall the concept later, because they made connections to other concepts previously learned.

Before encouraging students to create concept webs independently, you can create some as a whole-class activity, so that you can demonstrate ways to categorize details and indicate relationships. You may also wish to encourage the creation of concept webs for some small-group activities. If multiple groups create their own concept webs about a concept, these can be posted and shared to compare organizational strategies.

**PRIMARY EXAMPLE: STUDENT WORK**

### Creating a concept map about fractions

A primary student recalled all the information after working with fractions for a few weeks.

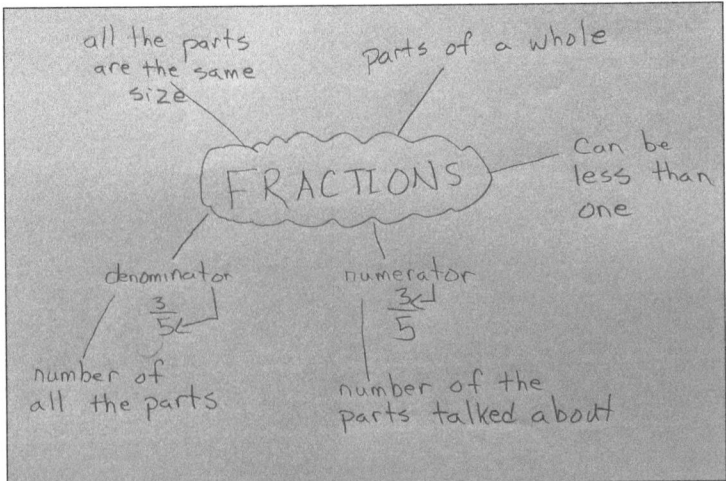

## Creating a concept web to explore triangles

The following concept map was developed by an elementary student. This student wanted to highlight triangles and associated key concepts: types of triangles, the sum of angles in a triangle, the number of sides in a triangle, and the straight edges that are characteristic of triangles.

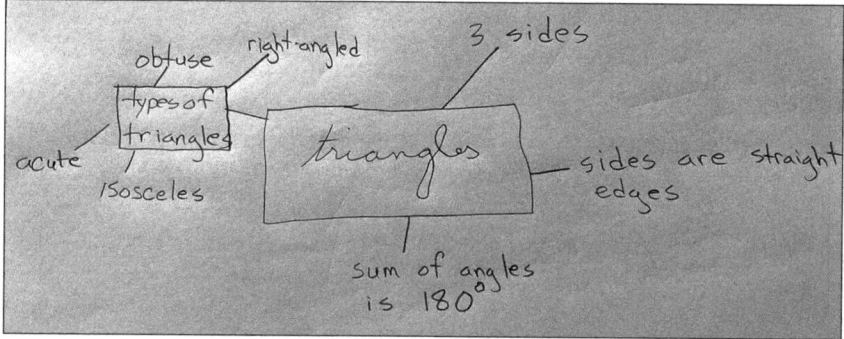

## Creating a concept map to explore order of operations

The following concept map was developed by an intermediate student. This student wanted to highlight order of operations, label the progression in BEDMAS, and include the different types of numbers for which order of operations could apply.

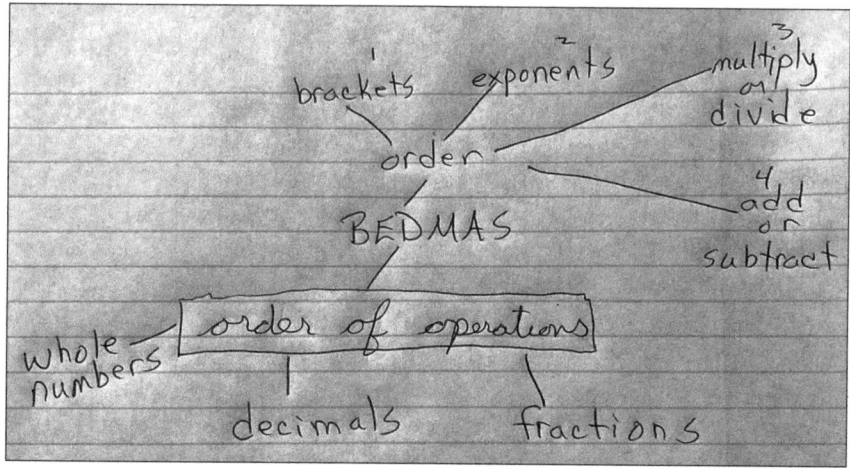

Depending on the age of students, visuals can be used instead of words to create the concept map. As students age, you can expect that they can include more text in the form of both terms and relationships.

# Students Create Graphic Organizers to Organize Thinking

**Suggested Prompts**

To inspire students as they create graphic organizers to organize thinking:

- *Which type of graphic organizer would help you sort through your thinking about the concept?*
- *What relationships are highlighted within this graphic organizer?*
- *How can you use this graphic organizer to help you think through a process?*

Sometimes, students need scaffolded support when they try to visualize concepts or see relationships amongst concepts.

Many students think of graphic organizers as a vehicle for recording information. What that misses, however, is the useful role that graphic organizers can play in the learning process. Graphic organizers are excellent tools for strengthening students' grasp of concepts by highlighting relationships, foundational details, and processes.

Graphic organizers are most useful if students think of them as a means to an end, and not the end itself. The act of sorting information into a graphic organizer can help the student organize and retain their learning of a concept. If you can emphasize this "why" of a graphic organizer—help students grasp the rationale for completing one—they will be in a better frame of mind to benefit from the exercise.

So, when using graphic organizers in the classroom, stress to your students that they should always begin with the process in mind: that the graphic organizer will help them identify relationships, highlight key details, and construct and retain knowledge. For graphic organizers to be an effective learning strategy, students must be aware that they are tools for learning, not a place to store information.

Following are four graphic organizers you may find helpful for assisting students in constructing knowledge, consolidating learning, and recalling previous learning.

## Using a Circles Graphic Organizer to Support Recall

Ask students to use this circles graphic organizer to recall prior knowledge related to a topic. They can begin by writing the topic. They then brainstorm concepts and ideas that relate to the topic in the inner circle. Then they choose the most pertinent information and write it in the outer circle.

*A circles graphic organizer*

**Using a circle graphic organizer to explore place value**

The following is a circles graphic organizer on place value completed by a primary student.

For graphic organizers to be an effective learning strategy, students must be aware that they are tools for learning, not a place to store information.

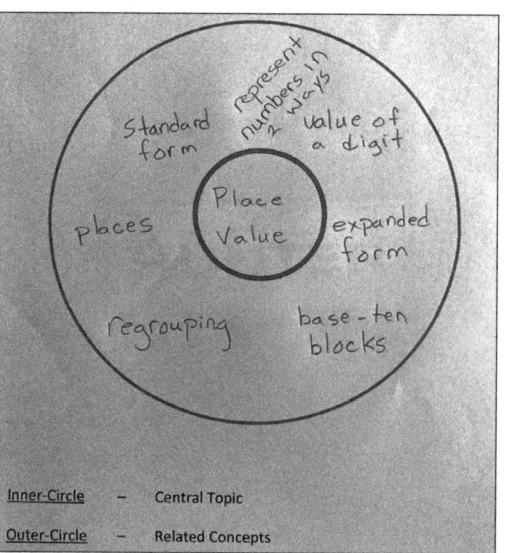

## Using an Organized-Circles Graphic Organizer to Recall and Organize Prior Knowledge

Ask students to use this organizer, like the previous one, to brainstorm concepts and ideas related to a topic. This organizer requires an additional layer of cognitive demand, as the related information is to be organized into four categories in the outer circle. You can offer the four categories before students begin brainstorming as a way to focus their efforts, or you can ask students to create the four categories themselves. This second option, where students create their own categories, adds an additional layer of difficulty to the learning experience.

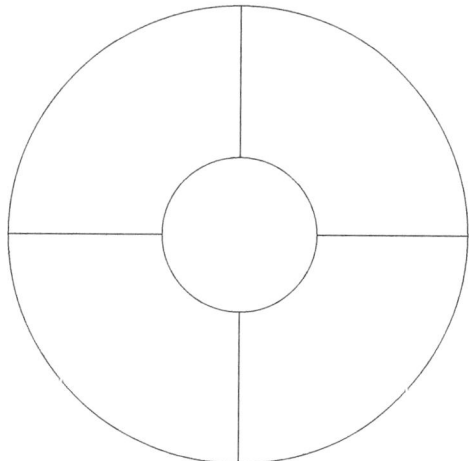

*An organized-circles graphic organizer*

## Using an organized-circles graphic organizer to explore probability

The following is an organized-circles graphic organizer on probability completed by an intermediate student.

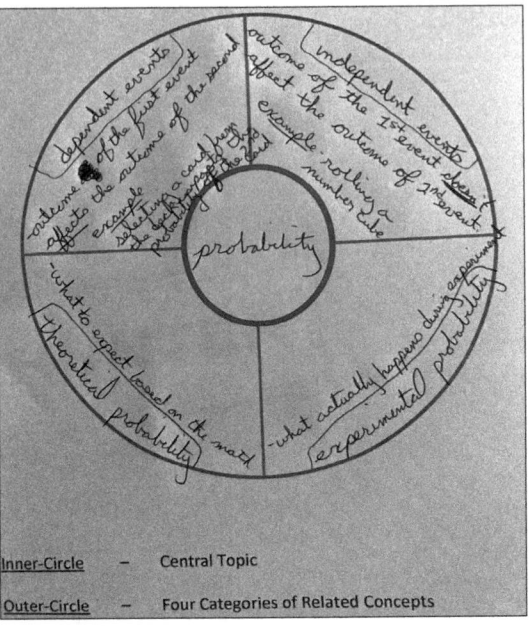

Inner-Circle  —  Central Topic

Outer-Circle  —  Four Categories of Related Concepts

## Using a Similar-but-Different Graphic Organizer to Compare and Contrast

The similar-but-different graphic organizer is like a Venn diagram but is laid out differently. Its purpose is to help students examine the characteristics (such as topics, concepts, and connections) of two things to distinguish which are different and which are similar.

Completing the similar-but-different graphic organizer adds layers of understanding to the learning and helps students develop their ability to recall information.

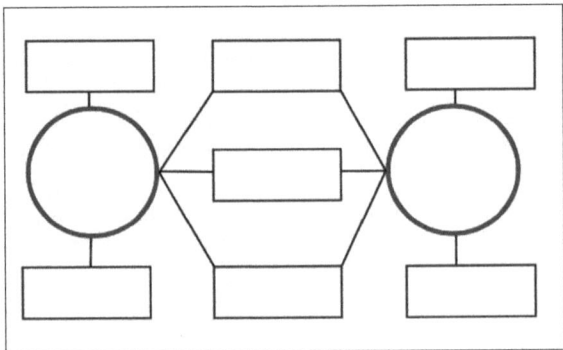

*A similar-but-different graphic organizer*

## Using a similar-but-different graphic organizer to compare triangles with trapezoids

The following visual duplicates a similar-but-different graphic organizer an elementary student completed to compare and contrast triangles and trapezoids. She began by writing the two items being compared in the circles. She then identified the characteristics that were different for triangle and trapezoid and recorded these in the outermost rectangles. Finally, she identified characteristics that were common to the two shapes and recorded these in the centre rectangles.

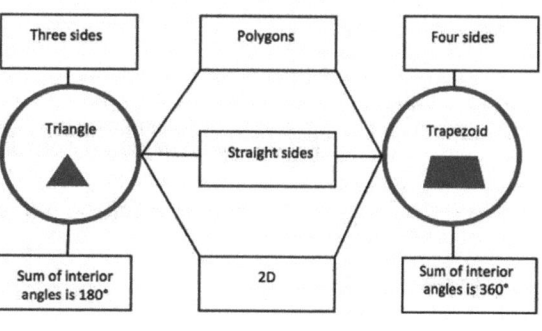

## Using a Tracking-a-Process Graphic Organizer to See Where Concepts Fall in a Process

The tracking-a-process graphic organizer is a visual to support students in understanding a process. Student understanding of a process can be strengthened by having them identify the sequence of steps through which a process occurs. The tracking-a-process graphic organizer can support students in doing that.

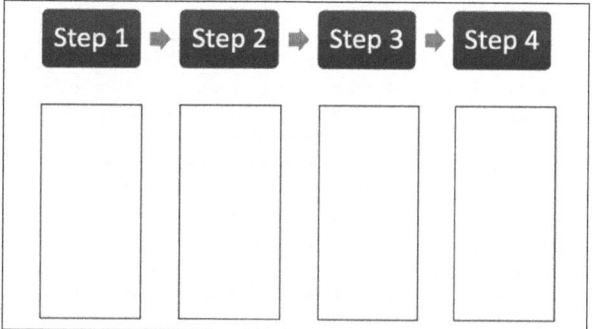

*A tracking-a-process graphic organizer*

Let's try the example of constructing a reflex angle. There are two processes students could follow to accurately construct a reflex angle. Using the graphic, student A devised these steps:

1) Draw a straight line with a dot in the centre, indicating a straight (or 180°) angle.
2) Determine how much larger the given angle is than 180°.

3) Draw the remaining degrees onto the straight angle drawn in step 1.
4) Draw a curved line around both angles.

Student B devised a different sequence of steps:

1) Calculate how much smaller a given angle is than 360°.
2) Draw an angle representing this difference.
3) Draw a curved line around the outside of this difference.
4) Measure and check.

Although different, each of these examples illustrate a correct use of tracking the process.

In the graphic organizer, students can add more detail for each step of the process and determine where details fall in this progression. Students can use both pictures and words to illustrate their understanding.

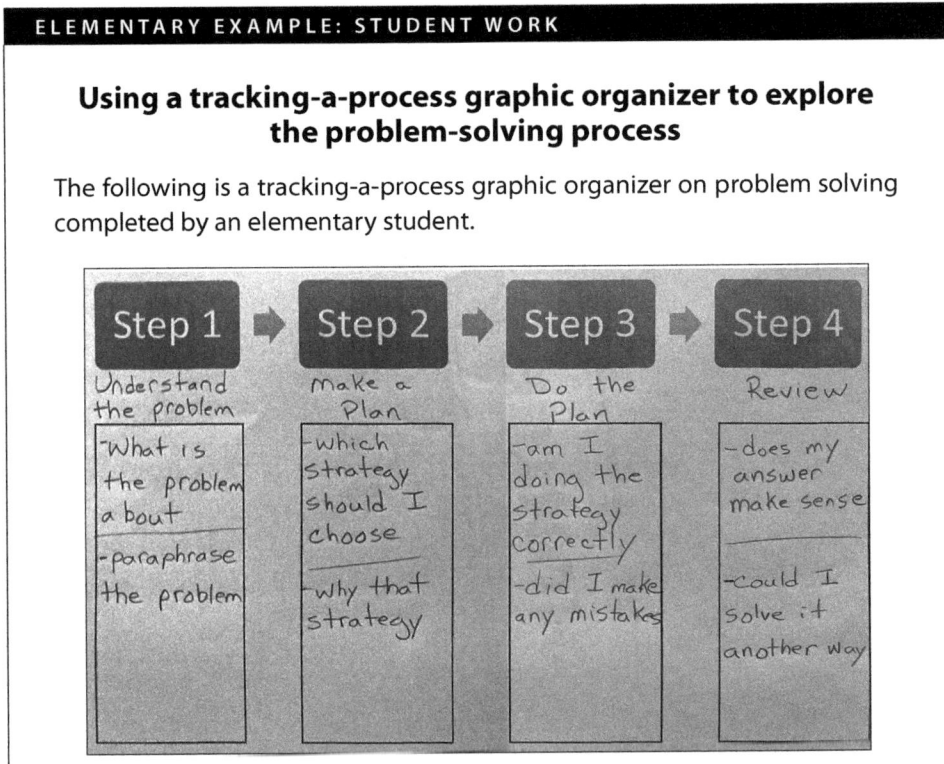

**ELEMENTARY EXAMPLE: STUDENT WORK**

### Using a tracking-a-process graphic organizer to explore the problem-solving process

The following is a tracking-a-process graphic organizer on problem solving completed by an elementary student.

| Step 1 | Step 2 | Step 3 | Step 4 |
|---|---|---|---|
| Understand the problem | Make a Plan | Do the Plan | Review |
| -What is the problem about<br><br>-paraphrase the problem | -which strategy should I choose<br><br>-why that strategy | -am I doing the strategy correctly<br><br>-did I make any mistakes | -does my answer make sense<br><br>-could I solve it another way |

## Learning by Picturing It Summary

Images play a significant role in student learning. Each of the strategies in Chapter 4—visualizing a problem to "see" it better, free sketching to explore problems, creating concept maps to find connections and relationships, and creating graphic organizers to organize thinking—are different ways of Learning by Picturing It. By giving your students the opportunity to use some of the strategies suggested, you will support them in building their ability to recall their learnings when necessary.

# 5

# Learning by Writing

*Help your students use informal, exploratory writing to help them think through mathematical concepts.*

How many times have you written a note to help you remember something? Or written down a question to ask, only to realize the answer shortly thereafter? Or written down an idea, only to revise it a few minutes later?

Human beings use the act of writing to clarify their thinking, remember important items, and work through problems. Generally, there are two types of writing: writing to demonstrate knowledge and writing to learn (Scardamalia & Bereiter, 1987). It is writing to learn that is the focus of this particular learning strategy. Writing is a powerful learning tool that helps us to consolidate our thinking and remember previous learning.

**Writing is a powerful learning tool that helps us to consolidate our thinking and remember previous learning.**

Writing to learn is an informal, exploratory approach to writing that consists of short, informal writing tasks that provide students support as they think through mathematical concepts. Writing-to-learn tasks should be low stakes, such that students will not worry about marks being prescribed to the writing. Students should feel free to poke around, experiment, and follow unfamiliar paths. The goal in writing-to-learn activities is to use writing as a tool to explore one's thinking, delve into concepts, and discover possibilities.

What does writing-to-learn look like? Usually it's messy, unorganized, and scattered; but these writings are also evidence of students thinking, making connections, and investigating possibilities. The goal is to help students learn, reflect on, and synthesize what they are learning. Writing-to-learn is a process whereby students strengthen previous learning, add layers of details to new learning, and make connections among concepts. It is a place for them to make their thinking visible to themselves.

**Suggested Prompts**

To inspire students as they write to learn:

- *What things do you understand?*
- *Write about what seems confusing!*
- *What questions do you have? Write them down! Can you think of more?*
- *Can you describe in writing that "aha!" moment you just had?*
- *What stumbling blocks have you encountered?*
- *What connections do you see between these two concepts?*

During writing-to-learn tasks, students are writing for themselves, to develop their mathematical understanding. They may decide to share a writing-to-learn piece, but the ultimate goal is to support learning, not demonstrate learning. Consider writing-to-learn tasks as a written form of a think aloud.

The right kind of feedback can make write-to-learn activities more useful. Students can go back to pieces that they have written earlier to answer any questions they asked themselves earlier or to clarify their thinking. Students can share their written pieces with you or their peers as a way to seek clarification on their thinking or to seek ideas for further exploration on what they have so far. Just be careful not to provide feedback related to spelling, punctuation, and so on.

To encourage consolidation and retrieval, provide students with multiple opportunities to write repeatedly on the same or similar topics over a period of time. This allows students to make connections while interrupting the process of forgetting. Further, encourage students to address different concepts from time to time so that they mix their review.

Depending on the age of the student or programming needs, the write-to-learn strategy can be modified to have students share their thinking orally while someone scribes for them. Using technology, students could make use of speech-to-text software to record their thinking in words.

---

**PRIMARY EXAMPLE: A STUDENT EXPLAINS**

### Writing to learn about patterns

We did a lot of work with patterns. That is how we started the year the last two years. I kept getting mixed up with what the repeating part of the pattern was. So I started to write down everything I knew about patterns in point form. I didn't care about periods or uppercase letters because that wasn't the point. The point was to get as much about patterns out of my head as possible. At first, when I did this, I couldn't write much. But every week when I did this I would write more and more. Now, after about a month, I remember a lot more about patterns.

---

**Building Retention**: This chapter introduces the following examples of Learning by Writing, which you can help your students use to build their retention capabilities:

- **paraphrase** a problem to help them understand it
- approach a problem by doing a **freewrite**
- do a freewrite of **30 words or less**
- use a **plus/minus chart** to record what they know and don't know

## Students Paraphrase a Problem to Help Them Understand It

The ability to paraphrase is an important skill to possess in math classrooms. To digest the content of a problem and articulate that understanding in one's own words is a significant aspect of encoding and consolidating. You can support students in rewording a problem, by both modelling and facilitating discussions with the class.

**Suggested Prompts**

To inspire students as they paraphrase a problem to help them understand it:

- *What are the main points within the problem?*
- *What are you being asked to do in the problem?*
- *Can you tell me what is going on in the problem?*
- *What words don't you understand that you believe are important in solving the problem?*

**How can we expect students to solve a problem if they cannot state it in their own words?**

When students paraphrase a problem successfully, you have a clear indicator that they understand the problem. Entailed in paraphrasing is the ability to review the content of the problem, identify the highlights of the problem, and then to rephrase in one's own words. Multiple steps are involved in paraphrasing; the student must understand various aspects of the problem and keep the original meaning when relaying the information to others.

In most cases, we can't. An inability to paraphrase may indicate a lack of understanding. And what is the point of students moving forward with a problem if they don't understand it? What will they use as a barometer of success? While asking students to paraphrase may help you assess students' level of understanding, more importantly, it may help students make sense of the problem in the first place.

Paraphrasing can consist of restating of the problem either orally or in writing. What is important is that the student identifies the highlights of the problem, shares what is being asked, and explains the context, including key details. Feedback can help them decide if their paraphrasing of the problem is accurate.

While attempting to understand and restate a problem in their own words, students may try to connect it to their lived experiences, compare it with other problems they have tackled, and begin to think of possible strategies they could use to work with the data. It is through this process of searching their previous learning and recalling it for the purpose of making meaning that students consolidate previous learnings and build new ones. Spacing and mixing paraphrasing a particular type of problem solidifies a students' long-term retention capability.

---

**PRIMARY EXAMPLE: STUDENT WORK**

### Paraphrasing part-part-whole words problems

After reading each word problem, a student tried to say it in her own words. This process helped her clarify what was missing and what was being asked.

---

**ELEMENTARY EXAMPLE: STUDENT WORK**

### Paraphrasing definitions

Math vocabulary can be challenging for students to understand. An elementary student recorded all the terms and meanings. Then she stated what the term meant in her own words. This process helped her own the learning, and it was a great opportunity to check for understanding and recall.

---

**INTERMEDIATE EXAMPLE: STUDENT WORK**

### Paraphrasing geometry problems

Intermediate students encounter a wide variety of formula for solving geometry problems. After a student learned about Pythagoras' theorem, she was challenged to complete some word problems using the theorem. She decided to explain each problem in her own words to figure out what it was asking and to make sure that she was using the formula correctly. This process helped consolidate her understanding of the theorem.

# Students Approach a Problem by Doing a Freewrite

**Suggested Prompts**

To inspire students as they approach a problem by doing a freewrite:

- *Write everything you know about this concept. You have 10 minutes!*
- *To get going, just write one thing you know about this concept.*
- *How did you decide what to write?*
- *You wrote a lot about this concept! What did you recall that can help you solve the problem?*

**DIGITAL SOLUTIONS**

Encourage students to record their thinking digitally using word processing software.

During a freewrite—a free-form retrieval—you ask students to write down anything and everything they know about a concept. Typically, you would give students a limited amount of time, perhaps 5, 10, or 15 minutes. The time you choose depends on the age of the student, their writing stamina, and the complexity of the concept.

A freewrite is an opportunity for students to recall previously learned concepts associated with a writing prompt, such as a word problem. The writing serves as a retrieval exercise that strengthens the pathways to the retrieved information and consolidates their memory of the concept.

During a freewrite, students have the opportunity to brainstorm about a topic without judgment. The idea is to generate as much information as possible related to the topic. From this, students can organize their thinking and revise their thoughts, all the while strengthening their connections to this knowledge. In addition, students may be informally connecting concepts as they search for any knowledge related to the concept being explored.

A freewrite is a flexible learning tool. You can use it within or immediately following a lesson, a day after the lesson, or after an extended period of time. Repeated use strengthens student memory.

The following three examples are freewrite samples taken from differing grade level bands. In each example, the teacher had modelled the freewrite strategy and provided various timeframes for the exercise.

---

**PRIMARY EXAMPLE: STUDENT WORK**

## Freewriting about multiplication

A primary student wrote the following freewrite about multiplication:

*Multiplication is about finding the number of items when you have some items in different groups. Each group needs to have the same number of items. This is why multiplication can be called equal groups. You can also show it with an array or on a number line. The answer to a multiplication problem is called product.*

---

**ELEMENTARY EXAMPLE: STUDENT WORK**

## Freewriting about transformations

An elementary student wrote the following freewrite about transformations:

*Translations, rotations, and reflections are three types of transformation. What you need is 2-D shapes so that you can do the transformations to it. Translations are when the 2-D shapes slide up, down, left, or right. Rotations are when the 2-D shape rotates on one of the corners. Reflections is when the 2-D shape is mirrored on a line of symmetry. You can do one, two, or three transformations for any 2-D shape.*

---

### Freewriting about surface area

An intermediate student wrote the following freewrite about surface area:

*Surface area is the total area of all the faces of a 3-D solid. You only look at the outside surfaces and you ignore anything on the inside. You have to find the area of each outer face and then add them all together to find the surface area.*

## Students Do a Freewrite of 30 Words or Less

To inspire students as they do a freewrite of 30 words or less:

- *You only have 30 words. Choose carefully!*
- *How did you decide what to include in your writing?*
- *Is there anything that you left out that you think you should have included?*

**DIGITAL SOLUTIONS**

Challenge students to write about a concept using fewer than 280 characters: the length of a tweet.

Like a freewrite, 30-words-or-less is an opportunity to write about a concept in a limited amount of time. The only difference is that 30-words-or-less has a cap on the total number of words. By introducing the cap, students need to be more focused on choosing the most crucial information from memory. Along with this, students need to be concise in what they write and to structure the information in an organized format so as to avoid exceeding the word count cap.

You can engage students with a 30-words-or-less challenge at any point in their learning. It is meant to support them in recalling and strengthening their understanding of a concept, so you can use the strategy before, during, after, or long after initial introduction of a concept.

You may wish to adjust the parameters of a 30-words-or-less challenge, depending on a student's age and writing ability. The cap could be extended to 40, 50, or 100 words. You could also adjust the time limit.

### Freewriting about estimation in 30 words or less

A primary student wrote the following freewrite about estimation:

*Estimation is about finding a close enough amount without having to count. I think of what 10 looks like and then wonder how many 10s are in the collection.*

### Freewriting about least common multiples in 30 words or less

An elementary student wrote the following freewrite about least common multiples:

*The least common multiple is the smallest multiple that each number has in common.*

**Freewriting about circumference in 30 words or less**

An intermediate student wrote the following freewrite about circumference:

*Circumference is the distance around a 2-D shape that is round.*

## Students Use a Plus/Minus Chart to Record What They Know and Don't Know

**Suggested Prompts**

To inspire students as they use a plus/minus chart to record what they know and don't know:

- *What do you know about the concept? Think about what we talked about in class discussion yesterday.*
- *What don't you yet know about the concept? Think about what you have difficulty doing.*
- *How do you know you are filling out the chart accurately?*
- *What could you do to move items from the minus column to the plus column?*

A plus/minus chart is a quick way to record learning. To fill it out, students first assess their thinking and then document their assessment in writing. The plus column is for items that they know well, and the minus column is for items that they do not yet know. By recording items in the appropriate column, students will steer their immediate learning goals.

You can generate a whole-class plus/minus chart or support students in crafting their individual charts. Encourage students to treat the plus/minus chart as a live document. From time to time, students can revisit their charts to review items in the minus column. Do they know enough about any of these items to move them to the plus column?

The plus/minus chart offers great flexibility in that students can use it to focus on one concept or multiple concepts. When using it to assess understanding of one concept, students would need to break it down to foundational items and then sort the items into the appropriate column. When using it to assess understanding of multiple concepts, students would list the concepts they know well under the plus and the concepts they do not yet know under the minus.

Student reviews using plus/minus charts (and concepts) can be mixed and spaced out over time and can therefore reinforce learning.

**Using a plus/minus chart to assess word-problem skills**

A primary student explores her thinking when solving word problems.

| PLUS | MINUS |
|---|---|
| - making a plan | - only reading the word problem 1 time |
| - writing a sentence for my answer | - not always understanding the question |
| - double checking my work | |

## Using a plus/minus chart to assess word-problem skills

An elementary student explores his thinking when solving word problems.

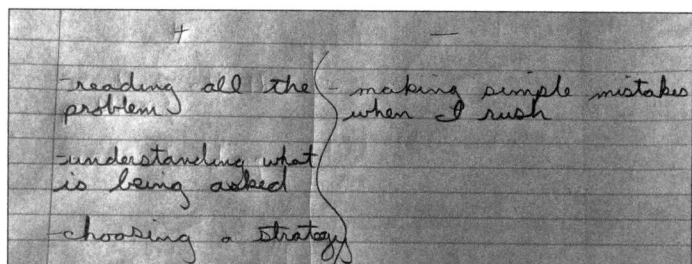

## Using a plus/minus chart to assess word-problem skills

An intermediate student explores her thinking when solving word problems.

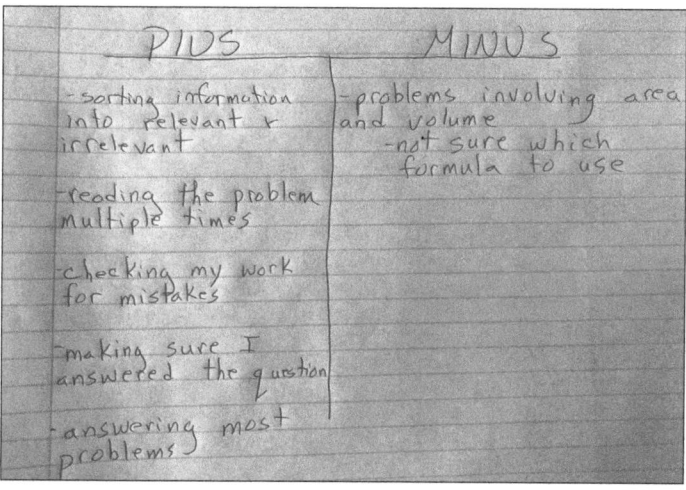

## Learning by Writing Summary

Each of the strategies in Chapter 5—paraphrasing problems to understand them better, approaching problems by doing a freewrite, doing a freewrite of 30 words or less, and using a plus/minus chart—are different ways of Learning by Writing. Each of these strategies involves informal writing to support learning. Students can go back and forth between the various strategies, depending on which concept they are exploring and which strategy is most helpful to them individually. It is not that one example is better than others; it is about finding the example that best supports the individual student.

**CHAPTER 6**

# Using Awareness Strategies to Improve Learning

*Help your students consciously improve their learning process.*

**Mathematics is not just about doing; it's about thinking.**

Have you ever read a book or taken a seminar only to realize at the end of it that you don't really understand much of the material? This happens all too often. As educators, we can support students in making sense of and retaining what they read and what they hear in class. Awareness strategies can play a significant role in helping students learn, understand how they learn, remember what they have learned, and apply what they have learned to new situations.

When discussing awareness strategies, I like to stress the notion that mathematics is about thinking, not just doing. Awareness strategies can improve how students approach a problem, work through it, and bring it to conclusion. They can affect the thinking that students do after solving a problem that will support them in developing their mathematical understanding. Far too frequently, people think of success in mathematics as solving numerous questions and getting them right. The math questions, however, are just tools for developing mathematical understanding. It's the understanding that is the goal. Awareness strategies are the engine that can build that understanding.

Marie M. Clay investigated the importance of awareness for reading generally. As people read a text, they problem solve by using strategies to make meaning (Clay, 1991). Being able to initiate or apply strategies at various times makes the reader more able to read independently.

Clay's conclusion applies equally to mathematics. If students apply awareness strategies when they engage with mathematical concepts, they strengthen their ability to make meaning, thereby layering the memory and increasing the opportunities for retrieval. Applying awareness strategies over a period of time and with a variety of concepts add to their prior knowledge for future work. Each experience goes into the students' math toolbox—something that they can access easily whenever they engage in mathematics.

Awareness strategies are flexible—they can be applied before, during, or after a problem.

- Before tackling a problem, awareness strategies can help students understand what the problem is about and what is being asked. They support students in deciding on a plan for successfully completing the problem.
- During the problem, students can apply awareness strategies to work through their plan and to monitor their understanding. It is the compass for them as they work.
- After the problem, students still need to take time to reflect on how their plan of action worked and what they can learn from their experience.

During reflection, students can analyze how they approached and worked through the problem, the strengths and weaknesses of their plan, and how they may do things differently the next time.

Students often omit the post-completion reflection. They do not see the value of rehashing what they did. Reflection, however, offers several opportunities to develop student learning. During reflection, students can analyze how they approached and worked through the problem, the strengths and weaknesses of their plan, and how they may do things differently the next time.

Through spacing and mixing, students can engage in awareness strategies throughout a unit, term, or year. By building awareness of how they learn and where they are on their learning path, students can develop their abilities to retrieve previous learning.

---

**INTERMEDIATE EXAMPLE: A STUDENT EXPLAINS**

### Reflecting as a strategy for moving forward

I always thought doing reflections was a waste of time. It just seemed to be me telling about all the work I did. But now I reflect on my work differently. I changed my mind about reflections when we were working on solving equations. Instead of just repeating what I did when I solved equations, I started to think about what worked well, what didn't work well, and what I would do differently the next time. I realized that, next time, I will address fractions first instead of adding and subtracting. It is kind of like critiquing my work to make me better for the next time. When reflecting, sometimes I remember concepts that I forgot. This helps me see that I could have approached and solved the problem differently than I did. My toolbox of strategies for math gets bigger when I reflect because I have time to think about my learning and moving forward.

---

**Building Retention**: This chapter introduces the following examples of Using Awareness Strategies to Improve Learning, which you can help your students use to build their retention capabilities:

- **prioritize** aspects of a problem
- **summarize** a problem-solving experience
- **reflect** on a problem-solving experience
- use a **learning journal** to record key takeaways
- record their **next steps**

# Students Prioritize Aspects of a Problem

**Suggested Prompts**

To inspire students as they prioritize aspects of a problem:
- *Which part of the problem is irrelevant?*
- *Which part of the problem is crucial for finding the solution?*
- *What do you need to know to tackle this problem?*
- *Which part of the concept do you need to focus on moving forward?*

Students are presented with an abundance of information within any single subject, unit, or lesson. They have an advantage if they are able to sift through this information and identify aspects that are significant to their understanding. Prioritizing is an awareness strategy that can help students impress in their memories the most important aspects of a problem, unit, or concept that they need to remember. They focus on the aspects that are necessary for them to have success with a concept and be able to apply it successfully in future.

What does prioritizing look like in the classroom? Perhaps you give students the opportunity to look back over a concept and highlight the aspects they believe are foundational to their personal understanding of the concept. Prioritizing can be done by students at the end of a lesson or unit, or when provided opportunities to recall concepts learned earlier in the school year.

## PRIMARY EXAMPLE: STUDENT WORK

### Prioritizing the key aspects of measurement

A primary student reflected on the measurement concepts he had learned about in the fall term: perimeter, comparing objects by certain attributes, using rulers to measure, and so on. He threw away all of the work samples and handouts that he didn't think were important to remember. He prioritized and kept only the materials that would help him remember the important aspects of these concepts.

## ELEMENTARY EXAMPLE: STUDENT WORK

### Prioritizing vocabulary that needs particular attention

An elementary student shared that she had a lot of math vocabulary to remember about angles. What she did was prioritize the vocabulary that she knew she was having difficulty with (reflex angle, obtuse angle, quadrant). Using this list, she could focus on her weakest vocabulary as she studied.

## INTERMEDIATE EXAMPLE: STUDENT WORK

### Prioritizing study of concepts most relevant to future study

An intermediate student reviewed all the concepts that she had studied during the previous term. She knew that some of these concepts would be important to help her understand material in the remainder of the school year. She identified these carryover concepts and prioritized her review on those. This helped her manage her studying and supported her in understanding and recalling these concepts later in the year.

# Students Summarize a Problem-Solving Experience

**Suggested Prompts**

To inspire students as they summarize a problem-solving experience:

- *What are the main points within the concept?*
- *What were the main thinking points during your problem-solving experience?*
- *In two sentences, tell me what you learned while doing this problem.*

Summarizing is the process of putting together information. As an awareness strategy, it involves reflecting on a learning experience (either during or after the experience), highlighting aspects that are significant to the learning, and concisely capturing it without having to restate every data point. To effectively summarize, the student must be able to distinguish between relevant and irrelevant information.

To summarize skillfully is helpful to students regardless of the discipline area. It requires an ability to sort through the text or problem to identify relevant information while disregarding irrelevant information. Not being distracted or taken off task displays determination and cognitive ability. The act of summarizing helps students to remember key points and to articulate these both while engaging with a problem and afterward. In addition, the resulting written summary may be useful for future reference.

Summarizing is a common practice in many literacy classrooms, but it also has a place in strengthening mathematical understanding. Students can look back to the day's lesson, a previous lesson, or a problem-solving experience that they may have engaged in. By summarizing that learning experience, students consolidate their learning and add opportunities for retrieval.

Asking students to summarize a group activity, problem type, or lesson for homework can be useful. But you might also like to try a strategy called stop-and-summarize, whereby you call out "Stop and summarize!" and students pause during the lesson and summarize their progress thus far. You could extend this to call out "Stop and summarize!" to prompt students to summarize what they have learned about a concept so far over a longer period of time.

The following three examples highlight how students summarized a problem-solving experience to highlight only those aspects that will support learning and application in future scenarios.

---

**PRIMARY EXAMPLE: STUDENT WORK**

### Summarizing a lesson

After a lesson, a primary student listed the main points of the lesson to his peers in a small group during consolidation. This act of summarizing gave the student a chance to recall the highlights of the lesson and to put these together in a concise and coherent manner.

---

**ELEMENTARY EXAMPLE: STUDENT WORK**

### Summarizing favored strategies

A student looked through multiple samples of their own work and listed the strategies they typically applied to solve problems of a particular type. Compiling this summary made them more aware of the strategies they gravitated to and strengthened their recall of these strategies.

---

**Summarizing the main aspects of favored strategies**

A student studied by looking back at previous work samples and summarizing each of the strategies they applied to solve various types of problems. These summaries helped the student recall strategies and to highlight their main aspects. This spaced approach to studying helped the student recall the strategies later.

## Students Reflect on a Problem-Solving Experience

**Suggested Prompts**

To inspire students as they reflect on a problem-solving experience:
- *What part of your problem-solving process went well?*
- *Pretend you're starting over. What would you do differently?*
- *What advice would you give to a student trying out this type of problem the first time?*
- *What part of the concept do you need to focus on more deeply?*

Reflection often gets pushed to the side, being considered a waste of precious classroom time. In fact, it could be the most valuable part of a lesson.

Reflection is a form of metacognition by which students examine what they know, are not quite sure of, and don't know yet. Reflection can give students an accurate understanding of their learning, which they can then use to plan for moving forward (Small, 2013). Reflection is a form of retrieval practice in that the student must think back and recall a learning experience, whether it was earlier in the day, week, or term. Through this retrieval, students strengthen the pathways to previous learning.

Reflection also provides students with the opportunity to elaborate on previous learning. By reflecting, students add layers of meaning to their learning. An example of this layering would be students asking themselves the following questions during reflection (Brown, Roediger III, & McDaniel, 2014):

- What went well?
- Would I have changed anything?
- What other knowledge or experience does this learning remind me of?
- What other strategies could I use to make my learning better?
- Is there anything I still need to learn to understand this concept better?

You can encourage the habit of reflection by providing specific time for private reflection. Conventionally, this occurs after major tasks, but you could also allow time at the beginning of class for students to reflect on their learning from the previous day. Prompt students with questions to think about, if necessary. They can record their thoughts in a journal or as an exit ticket.

The following three examples are reflections taken after students had the opportunity to revisit a concept after some time. You will notice how this reflective exercise shed light on some illusions of mastery that students had. By uncovering illusions of mastery, students will be aware of aspects they need to focus on in future. This thinking about their thinking strengthens students' understanding of a concept, as well as their ability to recall it in the future.

## Reflecting on what was successful

A teacher wanted to help a discouraged primary student. She suggested that the student refocus her reflections on what went well during math sessions. Doing this helped the student focus on the positive. She began to use what went well to solve other problems—and it helped!

## Reflecting to find ways to improve

A student was usually successful in math and did not see the point of reflection. The teacher challenged the student to reflect on how the problems could have been handled differently. So the student began to focus on how he could have changed his approach. By asking himself what he would change, the student soon realized that he could approach a problem from more than one angle and still be successful. He was more informed the next time he attempted to solve a similar problem.

## Reflecting on connections

At the conclusion of a problem-solving experience, a teacher asked students to reflect on what other knowledge or problem this experience reminded them of. This reflective approach helped students examine the learning for linkages with other concepts, experiences, or real-world connections. This strategy helped students consolidate their memory of the concept.

## Students Use a Learning Journal to Record Key Takeaways

**Suggested Prompts**

To inspire students as they use a learning journal to record key takeaways:

- *You can say anything you want.*
- *What problems are you having? What can you do well?*
- *Do you go back to previous entries to see if things have changed?*
- *Today, take a little time to look for a trend in your learning journal. Is it a trend you want to continue? What can you do about it?*

Learning journals are a place where students can "talk freely" about how things are going in math class. They can write about what is going well, what isn't going well, and how it all feels. The learning journal is a place not for practicing math but for telling about the experience of math. Students' interactions with concepts may appear in their journal as solutions, mistakes, questions, stumbling blocks, and "aha!" moments. Students can use their journals to record their thinking tracks through problems, to note difficulties or achievements, and to direct their personal learning journey.

To make the learning journal into a tool for retaining learning, encourage students to focus on their learning, not to worry about spelling or punctuation, and to keep their entries brief. Discourage long summaries or extended reflection. Have students fill their journals with multiple brief entries that record the steps in their math journeys, and which can therefore trigger memories. Guide students in making entries that are concise representations of learning events expressed in as few words as possible. Writing entries from such a perspective

**DIGITAL SOLUTIONS:**

In schools with a Bring Your Own Device (BYOD) policy, students can use their phones to keep their learning journals. This will provide them with easy access that may encourage use.

allows the student to recall an event quickly, and to reflect on it to move their thinking forward.

When students maintain a learning journal throughout a course, the journal provides a history of the students' thinking experiences. Students can review their personal histories to identify strengths and weaknesses in how they learn, knowledge that will support them with their metacognition. You can direct students to return to review the entries for a particular concept or unit at various times. In this way, you can space and mix review of concepts.

By maintaining a learning journal, students can develop a reflective perspective of their mathematical understanding. Students will notice the successes and struggles they experience over time and can use this knowledge to support their future encounters with novel situations. A handy method for developing such a reflective approach is to have students keep their learning journal in a digital format. Digital learning journals are appealing to many students, and they facilitate quick searches, making it easy to locate entries related to specific concepts, solutions, stumbling blocks, "aha!" moments, question types, and so on.

---

**PRIMARY EXAMPLE: STUDENT WORK**

### Recording strategies for regrouping

A primary student records a strategy she identified for knowing when to regroup:

*I can add two 3-digit numbers. I used to have problems with regrouping but I can do it now. I just have to ask myself first if I need to regroup and then I look through the numbers before starting to add. This gets me ready to know when to regroup.*

---

**ELEMENTARY EXAMPLE: STUDENT WORK**

### Recording difficulty with fractions

An elementary student records what part of a task he found difficult:

*I don't understand how to shift improper fractions to mixed numbers. I keep forgetting to keep the same denominator for the leftover fraction.*

---

**INTERMEDIATE EXAMPLE: STUDENT WORK**

### Recording success with rational numbers

An intermediate student records the path she took to success ordering rational numbers:

*Rational numbers can be hard to order. Lots of fractions, integers, and decimals. It took a while, but I can order all of them now. I was having trouble with fractions, but the more I practiced the better I got.*

# Students Record Their Next Steps

**Suggested Prompts**

To inspire students as they record their next steps:

- *What is the next step in the process?*
- *What part of the concept needs further work?*
- *How did you decide on your next step?*
- *How will you know when you have achieved your next step?*
- *Maybe that next step is a bit big. How could you break it down into smaller steps?*

Having easy access to their next-steps lists will allow students to constantly monitor their learning and to formulate their next area of focus.

Next steps is an informal writing activity that involves the recording of immediate short-term learning goals in relation to a single concept or multiple concepts. Students record their understanding of the topic and list the next step they must take to move their learning forward. These next-steps documents can take the form of an ongoing list of point-form statements, sentences, or formal writing.

Encourage students to keep their next-steps lists in a location that they can access repeatedly throughout the course of study—perhaps they can keep a digital copy that they can update easily. The next-steps list is a live document that students can revise as their learning progresses. They can treat it like a live to-do list, checking off items accomplished and adding new goals every day.

By having students maintain a next-steps list, you will be giving your students the opportunity to strengthen metacognition. They can use these lists to focus their work on areas they have identified as needing extra focus. By keeping students' minds on their goals, this strategy will add layers of memory to their understanding of a concept, which in turn will increase the likelihood of retrieval.

The following examples of next-steps entries highlight how this strategy can add details to memories and increase the ability of students to retrieve previous learning.

---

**PRIMARY EXAMPLE: STUDENT WORK**

### Setting a next step for multiplication

A primary student records a next step for getting better at multiplication:

*I can draw pictures for multiplication problems. I draw sticks as the things, and I draw circles around sticks to make groups. I am going to start to draw arrays for multiplication.*

---

**ELEMENTARY EXAMPLE: STUDENT WORK**

### Setting a next step for placing decimals on number lines

An elementary student records what she will try next to improve her work with number lines:

*I keep having trouble with placing decimals on number lines. I will use benchmarks on the number line to help figure out where to put decimals.*

---

**INTERMEDIATE EXAMPLE: STUDENT WORK**

### Setting a next step for dividing fractions

An intermediate student records a specific goal:

*I need to get better at dividing fractions using a formula.*

---

## Using Awareness Strategies to Improve Learning Summary

Awareness strategies engage students in being active in their own learning process. Each of the strategies in Chapter 6—prioritizing aspects of problems, summarizing problem-solving experiences, reflecting on those experiences, using a learning journal, and recording next steps—are different ways of Using Awareness Strategies to Improve Learning. They are all ways to help students construct meaning during the learning process. By introducing some of these strategies to your classroom, you will help your students both understand concepts deeply and recall their learning effectively.

# Part 1 Conclusion

Many students are encouraged to learn mathematical concepts using strategies that promote short-term performance. Cramming before a test, blocked practice, and rereading material (such as mathematics vocabulary) usually fail to help mathematical understanding stick. The confidence that this type of review provides is but an illusion of mastery.

In contrast, the strategies shared in the preceding chapters are intended to help you support students in all three stages of the learning process: encoding, consolidation, and retrieval. By encoding information, by consolidating information through the process of connecting concepts and layering details, and by retrieving memories of concepts previously learned, students develop their long-term retention and retrieval abilities.

You can introduce each of the learning strategies to the class as a whole and try them out together. You may zero in on class favorites, or individual students may choose strategies that work best for them.

The amount of time needed for any of these strategies is not much; in fact, the strategies are meant to take little time. An example of an intermingled use would be to do some dual coding (elaborating the concept by recording learnings in both visual and written forms) and then following up by creating a plus/minus chart to retrieve the learnings and identify gaps.

Remember that frame of mind is key to making these strategies work. Emphasize with students that you want them to accept the desirable difficulty of *forgetting* as a normal part of their learning process. The strategies that focus on self-assessment will help students be more aware of the gaps in their knowledge and be more aware of what strategies help them retain their learnings to fill those gaps. People want to know how to learn (Brown, Roediger III & McDaniel, 2014). The learning strategies provided in the preceding chapters will help your students learn how to learn and to understand themselves better as learners.

Remember that frame of mind is key to making these strategies work. Help your students to accept the desirable difficulty of *forgetting* as a normal part of their learning process.

# Instructional Strategies That Build Retention

How can you transform your classroom from a place for getting information into students' brains into a place where students experience all three stages of learning?

How can you transform your classroom from a place for getting information into students' brains into a place where students experience all three stages of learning: encoding, consolidation, and retrieval? So far, we have considered half of the answer: the learning strategies presented in Part 1. These strategies are ones that your students can use themselves to become more aware of and improve their own learning process. We explored the when, how, and why of them. By bringing these strategies into your classroom you will give your students the tools they need to take an active stance in their learning.

In Part 2, I would like to draw your attention to the second half of the answer: instructional strategies that seamlessly incorporate consolidation and retrieval of previously learned material into the classroom. In a sense, Part 1 was examining the challenge of improving the learning process from the students' perspective. Part 2 examines the challenge of improving the learning process from *your* perspective.

## A Collective Problem

It would be easy to assume that students forgetting previously learned materials is a problem experienced by individual students who do not apply themselves. Perhaps we simply need to teach better learning strategies such as the ones we explored in Part 1. But is it a problem of just certain individual students?

Before we can rectify the problem of students forgetting, perhaps we need to acknowledge that we have a larger problem on our hands. Too often I have heard teachers, coaches, and administrators comment on students not seeming to remember concepts that they have already learned *and demonstrated success with*. As a teacher, I have experienced this phenomenon in my classroom, regardless of the grade level I was teaching.

Consider the following testimonials, which three teachers have shared with me. They are representative of what I have heard countless times: teachers are concerned that large numbers of students are not able to recall previous learning. All three teachers express a concern that a majority of students were demonstrating success in previous grades or earlier in the year yet were quite unable to apply their learning later.

---

**PRIMARY EXAMPLE: A TEACHER EXPLAINS**

### Forgetting part-part-whole from year to year

A lot of time is spent on part-part-whole in Kindergarten, Grade 1, and Grade 2. I know for a fact that students work with this concept quite a bit and that they demonstrate understanding of it, whether a part is missing or the whole is missing.

Why, then, does it feel as though we have to introduce it again from scratch each year? I can see a little forgetting of part-part-whole, but for students to act as though it is a new concept, I just don't get it.

How nice would it be if we could hit the ground running from where students left off with the concept from the previous year?

It's not as if I don't think it is important to teach, I just wish I could build on their previous learning and that students would be able to remember the concept and their experiences with it.

---

**ELEMENTARY EXAMPLE: A TEACHER EXPLAINS**

### Forgetting how to order fractions from year to year

Fractions, fractions, fractions. I see the same thing happening every year. I introduce fractions to students—we represent them using concrete objects, pictures, and number lines. We need to do this well because it is important for students to have this understanding when they are asked to compare two fractions.

I first have students compare fractions with like denominators, then with like numerators. From there, I then move into having students compare two fractions with unlike numerators and denominators. I know that, in the past, they have spent considerable time on finding equivalent fractions to help them understand and use pictures and number lines.

However, when I recently gave these students two fractions with like numerators to compare, they completely forgot what they had demonstrated understanding of in the past (that is, using number lines and pictures) and placed an equal sign between the two fractions. When I gave students two fractions with unlike numerators and denominators, they said that they weren't sure and just guessed with the > sign.

Why are students forgetting what they did in earlier grades and earlier in the year? I thought they had this, and I even gave them grade level on the report card.

---

## Forgetting order of operations from year to year

I know that students work with BEDMAS, or order of operations, in the grades before mine. When they see it first, they don't work with exponents, but they get the gist of it. I have seen their work samples demonstrating this understanding. Many students, when engaging with problems requiring BEDMAS, say that it is easy.

When I start working with them, we focus on how to include exponents in the process. They seem to do well with this. A little confusion at the beginning, but over time they get the hang of it.

When moving into equations and solving for variables, it is like they haven't seen BEDMAS before. I'll see a countless number of work samples with students working left-to-right, ignoring the order of operations. As I look through their responses, I see the same trend: operations being done from left to right without any consideration for BEDMAS. Then, when I ask students to double-check their work, they don't see any error. They often just look to see if they completed the operations correctly, and, if they did, they say that they're right.

It's frustrating because I know that they can do it, but they don't do it when solving for an unknown variable. So I have to stop the lesson and go back to teach them how to do BEDMAS again. This is not working for them or me. It seems that my time spent working on BEDMAS earlier in the year is for naught.

**Many teachers face mass forgetting in their classrooms.**

Many teachers continue to face mass forgetting in their classrooms. The interesting part of each of the above cases is that at some point in the year or a previous year, a majority of students demonstrated understanding of the concepts. Multiple factors may be working together to create this problem. It may be unclear what factor is most responsible for the trend toward mass forgetting, but we do have a trend and we have to consider the role that instruction can play in changing the trend.

## The Failure of Traditional Review Practices

As you read the following testimonial, ask how many times you experience something similar, especially before a mid-year test or end-of-year standardized test.

## Attempting to review adding and subtracting fractions

An intermediate teacher was preparing his class for a standardized test at the end of the school year. The teacher decided to ask a few questions focusing on a topic addressed mid-year. As a class, students had done very well with this topic when assessed. The teacher thought this would be a quick one-day review, two days at maximum.

The concept was adding and subtracting fractions with unlike denominators.

The teacher planned to model two problems—one addition and one subtraction—on the whiteboard, so that students would understand the focus of the review.

This only took a few minutes, but many students in the classroom said that they had *forgotten* how to add and subtract fractions. Others couldn't even remember working on this concept earlier in the year.

What happened next was unsettling for the teacher:

- Approximately two-thirds of the class struggled to work through the questions, while the remaining one-third of the class raised their hands asking the teacher to *show* them how to solve the questions.
- The teacher went around the classroom, offering support and providing hints as to how to begin (for example, highlighting that the denominators were different and that equivalent fractions would be helpful).
- Eventually, the teacher paused the independent practice portion of the review and started recording answers to the questions onto the whiteboard as students recorded the solutions onto their review sheets.

Reflecting on this, the teacher decided to continue the review of adding and subtracting fractions with unlike denominators the next day.

Interestingly, on the following day, when the teacher started the next day's lesson with a subtraction question involving two fractions with like denominators, half of the students did the unnecessary step of finding equivalent fractions as their first step. What this showed is that students were not yet understanding addition and subtraction of fractions although they had reviewed the process just the day before.

**The term *review* should not be used to describe what is, essentially, cramming.**

A lesson we can take from the above scenario can be summed up with the following phrase: "Just because you may know something at one point doesn't mean you'll remember it."

During a typical period of review, teachers stop addressing new material so that they have time to review material previously addressed in the hopes that students will have it *fresh* in their mind for the test. Such an approach to instruction values short-term performance instead of meaningful interaction with the curriculum and standards. Is this form of review worthwhile for support of long-term learning? Or is it just a form of cramming for short-term performance? Using the term *review* to describe this approach to instruction and student learning is both misleading and inaccurate.

**How can we provide our students with more than just superficial learning and short-term performance?**

The typical form of review is symptomatic of instructional approaches that do not promote meaningful student learning. It is not enough to approach instruction from a one-and-done perspective. So how can we change the situation in our classrooms? How can we provide our students with more than just superficial learning and short-term performance?

A good place to start is to consider the instructional strategies presented in Part 2. These strategies can be tools in your toolbox, designed to help you teach concepts from the perspective of the broadened learning definition introduced in the Introduction and Part 1. By using these strategies, you can enable your students to encode learnings, strengthen their understandings, and recall previous learning to solve future problems.

## Instructional Strategies that Support Long-Term Learning

The chapters within Part 2 present instructional strategies that have the potential to benefit your entire class. They will enable you to better use monitoring, year planning, classroom review, testing, and homework to help your students consolidate and retain what they have learned. While some of the strategies may seem commonplace, most significant is how each strategy is applied. All the strategies here are about doing things differently, not doing more things. As stated earlier in this book, the goal is not to work harder but to work smarter.

The table on the next page highlights the instructional strategies you can apply to strengthen student understanding and recall of previous learning. In the indicated chapters, you will find suggestions for leveraging these instructional strategies within classroom routines. Each of the strategies can work seamlessly within the various instructional models that you currently employ.

## Instructional Strategies

| Strategies | In the Classroom | How It Works |
|---|---|---|
| **Chapter 7: Expanding Use of Formative Assessment** | Shift your assessment efforts toward ongoing formative assessment. | **Use formative assessment frequently and consistently** to gain an informed discernment of students' mathematical understanding. Varied use of conversation, observation, and assessment of product together provide a comprehensive picture of what practice and review students need. |
| **Chapter 8: Focusing Instruction on Key Concepts** | Prioritize concepts that best support student learning across mathematics. | Key concepts are those curriculum outcomes and standards that have significant impact on the learning of other concepts. By identifying key concepts, you can craft instruction that helps math concepts stick in student understanding. This chapter introduces ways that you can **return to key concepts repeatedly** throughout the year, thereby supporting long-term retention. |
| **Chapter 9: Rethinking Year Plans** | Plan the year to interrupt the forgetting process. | Year plans are the structure of a school year. You can **craft year plans in a way that promotes retention** of concepts and supports students in retrieving their learning for future applications. The examples shared in this chapter will help you structure a school year in a way that increases the impact of your instruction. |
| **Chapter 10: Rethinking Review** | Create spaced and mixed reviews to interrupt the forgetting process. | Review is one of the most misunderstood strategies in education. You can **create purposeful, effective review by spacing and mixing**, so that students have ongoing opportunities to engage with concepts and apply them to solve problems. The examples shared within this chapter can be adapted easily to all grade levels and will provide students with opportunities to strengthen their learning and ability to recall this learning. |
| **Chapter 11: Rethinking Testing** | Use frequent testing to spark more frequent retrieval. | Testing is often misunderstood to be simply a summative assessment tool. Testing can, in fact, serve as a powerful instructional strategy when applied appropriately. This chapter highlights two approaches to help you **use testing to support learning and retention**. |
| **Chapter 12: Rethinking Homework** | Reimagine homework as an opportunity for review. | Homework can play a role in supporting students as they learn mathematical concepts. Through purposeful planning and effective structures, you can **craft homework that gives students meaningful opportunities to practice concepts**. |
| **Chapter 13: Adjusting Common Classroom Tools and Strategies** | Apply a retrieval practice approach to common instructional tools and strategies. | You can tweak countless classroom tools and strategies to **have a positive impact on student learning**. The examples shared in this chapter can be added easily to any classroom to support a spaced and mixed approach to instruction. |

# 7

# Expanding Use of Formative Assessment

*Shift your assessment efforts toward ongoing formative assessment.*

We all know that assessment is important. If there is quality evidence of learning available to the teacher, decisions about student learning can be made based on reliable and valid information (Davies, 2011; Small, 2013). Too often, however, teachers rely primarily on summative assessment—largely end-of-unit or end-of-term tests and projects.

**The focus on summative assessment has a huge drawback. It provides a single snapshot of short-term performance.**

The focus on summative assessment has a huge drawback. It provides a single snapshot of short-term performance. To see the big picture—and to see evidence of students retrieving learning multiple times in multiple ways—you cannot limit your evidence in this way.

I recommend collecting evidence of learning on an ongoing basis for three reasons.

- First, you can monitor student progress over time. This gives you a better understanding of where students are at—what they can recall and what they cannot recall. By having this information, you know which previous learnings students need help remembering.
- Second, the assessment process itself can offer multiple opportunities for students to consolidate and retrieve learning. This remembering process is crucial for getting learnings into long-term memory.
- Third, the assessment process can provide students with feedback, which can help them know where to focus future study efforts.

Your sources of evidence for formative assessment can have huge impact on learning. The key is to collect evidence in a variety of ways. I suggest a triangulation approach to collection, an approach that rests on three sources of evidence: conversation, observation, and product (Davies, 2011). The following diagram is a visual representation of this triad.

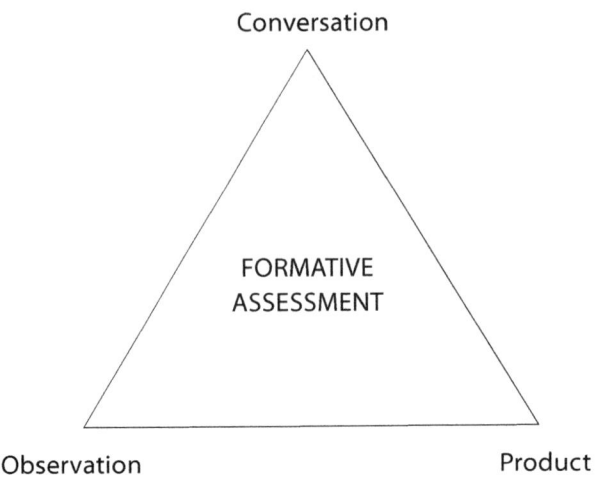

*The triad of formative assessment forms*

These three sources of evidence can provide you with a variety of information. If you talk with the student about the concept, observe the student as they apply the concept, and then see completed work samples, you will gain a well-rounded understanding of student understanding. At the same time, students gain the opportunity to engage with a concept from differing perspectives, which strengthens learning. Talking about a concept, for example, adds a layer of detail to the memory, thereby increasing the likelihood of recall.

## Gather Information through Conversation

**DIGITAL SOLUTIONS**
Use a digital platform to meet with students and record the conversation.

Conversations about student learning can involve two aspects: listening to what students are saying about their learning, and reading what they have documented about their learning (Davies, 2011). Conversations can be face to face or written. Included in conversational evidence of learning are conferences, one-to-one conversations, classroom discussions, and student self-assessments. Through these exchanges, you can gain insight into what the student thinks, both about the learning goal and about themselves as learners.

Conversations can be a valuable tool in supporting student retention and application of learning. Consider how impromptu conversations can assist students in recalling previous learning. For example, place value is often introduced early in the school year. Why not engage students in conversations addressing previous learning? These conversations allow you to assess how well students have retained the previous learning. Also, as students are engaged in the conversation, they are retrieving previous learning, thereby strengthening the specific understanding of the concept (consolidation) and strengthening the pathways to this memory. It is important that students have opportunities to converse about learning—through talk, they revise and reframe their understanding. Such revising and reframing creates a desirable difficulty as it involves the interruption of forgetting.

## Gather Information through Observation

During instruction and independent tasks, teachers are generally involved in some form of observation, consciously or subconsciously. Observational data can be collected when students are engaged with a concept, strategy, or process. Davies (2011) stresses that it is not enough to observe students as they complete a task; instead, teachers should also observe students while they are engaged with the task.

Observational data is most helpful when you plan how you will collect the data, what you will be looking for, and how you will document what you see. The task students are engaged in must closely relate to the concept, strategy, or process you have identified as the learning goal. Be sure you are collecting observational data at a time when students will provide you with evidence of their learning; observation is only beneficial when the task aligns with the learning goal.

It is one thing to talk with students about previous learning; it is quite another to watch students as they attempt to retrieve the previous learning independently.

I encourage you to employ observation when you ask students to engage in a task related to previous learning. It is one thing to talk with students about previous learning; it is quite another to watch students as they attempt to retrieve the previous learning independently. When students are required to work with a concept they have not experienced in some time, why not observe them as they enter the problem, work through it, and arrive at a conclusion? Through observation, you can determine the productive struggle the student experiences during the task, see how successfully they complete it, and decide what next steps you would like them to take.

## Gather Information by Assessing Product

Product is the source of evidence that many teachers are most comfortable with. More than likely, it is the source that was most used when we were students. Products are tangible objects that you can collect and look through for evidence of learning. Assessing product allows us to examine whether, and to what degree, a student can demonstrate understanding. Examples of product are homework, quizzes or tests, portfolios, tasks, and journals.

To support student retention and application of previous learning, I suggest including questions related to previous learning in virtually any assignment. Such an approach provides documentation of a student's ability to recall and apply that previous learning, sometimes long after you originally taught it. Your assessment of product then enables you to assess and address the ability of a student to retain previous learning and apply it to novel situations.

## Identifying Patterns and Trends

A question I hear from many teachers regards quantity of evidence: How much is enough? You can be confident that you have enough evidence when you can identify patterns and trends in student learning. By seeing the patterns and trends, you can understand the learning needs of your students, identify next steps, and provide feedback.

To identify patterns or trends, you must assess student learning in a variety of contexts over time. You might assess student recall through quizzes, group work, discussions, and so on. Sort these assessments into three categories: conversations, observations, and products.

- How well do students **converse** with you about the concept or strategy?
- How well do students engage with the concept or strategy while being **observed**?
- How well do students demonstrate understanding of the concept or strategy through a **product**?

Framing these assessment opportunities into these three categories (conversation, observation, and product) enables you to triangulate your findings. This triangulation can be further strengthened when you assess student learning over time.

The following examples highlight how three teachers applied conversation, observation, and product as methods to monitor students' ongoing progress.

---

**PRIMARY EXAMPLE: A TEACHER EXPLAINS**

### Assessing ability to solve word problems

It is important that I have a good grasp of student understanding when working on word problems. We do a lot of word problems in my class because it's important for students to know not only how to add and subtract but also when to add and subtract. I find that word problems give me both things.

I used to rely on giving students word problems on paper, one problem per sheet. I would then collect these from students and look through their work for evidence. What I found was that I was not getting a good picture of their thinking. I was just seeing the answer and a sentence, if I was lucky. I needed to know more about their thinking and about how they worked through the problem.

Now I still give students a word problem on paper for them to work on independently and submit; but I also sit down with students and have a conversation with them focusing on what the word problem is about, what strategy they will apply and why, and how they will know if their response is correct or not.

I also observe students as they work through a problem so that I can see how they approach the problem, when they change their mind during the problem-solving process, and how they know they have reached a conclusion.

I find that balancing these three ways to assessing thinking provides me with a more complete picture of student thinking. It also has made things a bit more manageable, as I am not collecting paper all the time.

---

**PRIMARY EXAMPLE: A TEACHER'S EXPERIENCE**

### Conferencing as an instructional strategy

A primary teacher shared how she met with students to discuss their understanding of a concept. The conversations were fluid in that they evolved depending on what students wanted to discuss about their thinking. The teacher found such conferences to be a great way to build student mathematical discourse and co-create learning goals. She especially liked to conference on concepts that hadn't been addressed in the classroom for some time. Doing so provided good data on how well the student recalled learning from earlier in the year.

### Observation as an instructional strategy

During independent practice, an elementary teacher would observe students as they were engaged with a mathematical task. The teacher especially paid attention when students were recalling previous learning. She watched closely as her students undertook the task and monitored how they worked through the process. It was important to determine stumbling blocks and points of clarity as students navigated the problem. The teacher noted what concrete materials the students used, what additional resources they referred to, and what questions they asked during the process.

### Assessing product over time

An intermediate teacher shared how she had been assessing through conversation and observation, but that she needed to collect work samples to determine if students could successfully solve a particular type of problem. The teacher collected a work sample, analyzed it for evidence of student thinking, and then added it to the student's learning portfolio. She repeated this process several times over several months. These work samples served as a documentation of the student's progression of learning. The work samples provided insight into whether the student was demonstrating growth in the ability to recall and apply a concept addressed earlier in the school year.

## Expanding Use of Formative Assessment Summary

To gain a complete picture of student progress, have conversations with students about their thinking, observe them as they are engaged in tasks, and look over some of their completed work samples. Assessment based on conversation, observation, and product can help you to fairly judge a student's ability to retain and apply previous learning. By assessing over time, you can determine if students have developed a long-term understanding of the concept or strategy. In addition, a spaced and mixed approach to assessment enhances student ability to retain and apply previous learning because students must work through the desirable difficulty of interrupting the process of forgetting.

**CHAPTER 8**

# Focusing Instruction on Key Concepts

*Prioritize concepts that best support student learning across mathematics.*

## Strategy: Continually Return to Key Concepts throughout the School Year

**DIGITAL SOLUTIONS**
Use an interactive whiteboard system to sort through and highlight key concepts within the curriculum.

When teachers attempt to digest the number and variety of curriculum outcomes and standards they need to address in a single year, they can get overwhelmed. There are so many! While all curriculum outcomes and standards are important in supporting student mathematical understanding, however, not all of them have the same impact on student learning. Some curriculum outcomes and standards have greater potential to add value—they are key concepts.

Key concepts are those curriculum outcomes and standards that allow other outcomes and standards to make sense and be learned more efficiently and effectively (Reeves, 2010). Key concepts come directly from the curriculum and standards documents. Key concepts form the basis for further learning, whether that learning takes place during the current school year or in subsequent years.

Key concepts can be described as the small group of concepts that can ensure ongoing success within and between grades, but only if students know them well. A focus on key concepts does not, and should not, preclude the instruction and assessment of all other required curriculum outcomes and standards. All outcomes and standards are important.

Labelling certain concepts as key is merely identifying the highest priority material. Think about it. We know that many of the mathematical concepts are interconnected and can be approached as supports for one another. One mathematical concept can support the understanding of another. So, when selecting curriculum outcomes and standards to be key concepts, one or more of the following criteria must be met (Reeves, 2010):

- **leverage**: outcomes and standards that focus on knowledge and skills used in multiple academic disciplines
- **endurance**: outcomes and standards that focus on knowledge and skills that will be relevant throughout a student's lifetime
- **essentiality**: outcomes and standards that focus on the knowledge and skills necessary for students to succeed in the next grade level

For clarity, consider the following examples.

- an outcome meeting the criteria of **leverage**: creating, labelling, and interpreting line graphs to draw conclusions
- an outcome meeting the criteria of **endurance**: using fact-learning strategies to solve problems
- an outcome meeting the criteria of **essentiality**: understanding area

While these examples highlight outcomes addressing a single criterion, some outcomes will address two or three of the criteria. Those are likely good choices to be key concepts.

The strategy of shifting toward formative assessment, as explored in the previous chapter, works hand in hand with the strategy of focusing on key concepts. How does this shift your year? Through the course of a year, you assess all outcomes and standards, as you normally would do. Ideally, you give students multiple opportunities to be active in the process. At the same time, though, you would schedule assessment opportunities for all your chosen key concepts throughout the year. These key concept checkpoints would help you assess for learning over time. In addition, you could provide assessment opportunities so that you can assess the key concepts in different contexts (conversation, observation, or product) to assess for student understanding within different venues. By providing more opportunities to assess for key concepts, you would be assessing for both the key concepts and other curriculum outcomes and standards.

Note that all grade level curriculum outcomes and standards must be addressed and worked with at some point during a school year. The key concepts strategy is not about instructing and learning only a few curriculum outcomes and standards during the year. Instead, it is about prioritizing outcomes and standards that will impact student learning across mathematics and other disciplines and providing support to students as they work with other outcomes and standards.

As students return to the key concepts over and over during a year, they add layers of detail to their memory of the concept. They also strengthen the pathways to this memory every time they access it. The key concept approach to curriculum and standards provides more opportunities for students to strengthen the memory and pathways to the memory, thereby supporting other concepts that are connected to them.

The key concepts strategy is not about instructing and learning only a few curriculum outcomes and standards during the year. Instead, it is about prioritizing the outcomes and standards that will most impact student learning.

## Working with multiplication facts as a key concept

It's just plain necessary that students know their multiplication facts if they are to succeed with multiplication and division. Students cannot be spending all their energy on solving facts when they are in the middle of a problem-solving scenario involving multiplication of two-digit numbers.

I believe that the multiplication facts are foundational for my students' math understanding, so I identified them as one of my must-haves for the year. What this means is that I keep coming back to them and assessing students on how they are progressing. It isn't enough that I just focus on them for a block of time and then move on. When I did this in the past, students forgot most of the facts and relied on skip counting to find all products higher than 20.

Now, I continuously bring fact learning up throughout the year, monitor student abilities on understanding and applying certain strategies, and have them make goals as to what to work on for the foreseeable future.

What I have seen, since adopting this system, is that fact learning has gone up in my class. Not only can students recall the multiplication facts up to 9 times 9, but they can apply that knowledge easily when solving multiplication and division problems. They are now focusing on the problem solving instead of skip counting to find products.

I really like this approach and I'm going to start identifying other must-haves for the year. That way, my students can have this success with other important concepts.

The following examples list key concepts chosen for three grade levels. Each of these concept lists were shared by teachers. Different teachers may select a different set of key concepts for their students depending on which curriculum they are tasked to work with, as well as their personal teaching style and interpretation of mathematics.

## List of key concepts

The following is a list of the key concepts a teacher crafted for a primary math class.

- Count to 200, including by 20s, 25s, and 50s, using a variety of tools and strategies.
- Read, represent, compose, and decompose whole numbers up to and including 200, using a variety of tools and strategies, and describe various ways they are used in everyday life.
- Compare and order whole numbers up to and including 200, in various contexts.
- Estimate the number of objects in collections of up to 200 and verify their estimates by counting.

- Recall and demonstrate addition facts for numbers up to 20, and related subtraction facts.
- Use objects, diagrams, and equations to represent, describe, and solve situations involving addition and subtraction of whole numbers that add up to no more than 100.
- Determine pattern rules and use them to extend patterns, make and justify predictions, and identify missing elements in patterns represented with shapes and numbers.
- Determine what needs to be added to or subtracted from addition and subtraction expressions to make them equivalent.

---

**ELEMENTARY EXAMPLE**

## List of key concepts

The following is a list of the key concepts a teacher crafted for an elementary math class.

- Read, write, and compare decimals to thousandths.
- Fluently multiply multi-digit whole numbers using the standard algorithm.
- Find whole-number quotients of whole numbers with up to four-digit dividends and two-digit divisors, using strategies based on place value, the properties of operations, and the relationship between multiplication and division. Illustrate and explain the calculation by using equations, rectangular arrays, and area models.
- Add, subtract, multiply, and divide decimals to hundredths, using concrete models or drawings and strategies based on place value, properties of operations, and the relationship between addition and subtraction; relate the strategy to a written method and explain the reasoning used.
- Add and subtract fractions with unlike denominators (including mixed numbers) by replacing given fractions with equivalent fractions in such a way as to produce an equivalent sum or difference of fractions with like denominators.
- Interpret a fraction as division of the numerator by the denominator ($\frac{a}{b} = a \div b$). Solve word problems involving division of whole numbers leading to answers in the form of fractions or mixed numbers, e.g., by using visual fraction models or equations to represent the problem.
- Recognize volume as an attribute of solid figures and understand concepts of volume measurement.

## List of key concepts

The following is a list of the key concepts a teacher crafted for an intermediate math class.

- Read, represent, compare, and order rational numbers, including positive and negative fractions and decimal numbers to thousandths, in various contexts.
- Use the properties and order of operations and the relationships between operations, to solve problems involving whole numbers, decimal numbers, fractions, ratios, rates, and percent, including those requiring multiple steps or multiple operations.
- Use objects, diagrams, and equations to represent, describe, and solve situations involving addition and subtraction of integers.
- Add and subtract fractions, including by creating equivalent fractions, in various contexts.
- Solve equations that involve multiple terms, whole numbers, and decimal numbers in various contexts, and verify solutions.
- Show the relationships among the radius, diameter, and area of a circle, and use these relationships to explain the formula for measuring the area of a circle and to solve related problems.
- Select from among a variety of graphs, including circle graphs, the type of graph best suited to represent various sets of data; display the data in the graphs with proper sources, titles, labels, and appropriate scales; and justify their choice of graphs.
- Determine the impact of adding or removing data from a data set on a measure of central tendency and describe how these changes alter the shape and distribution of the data.

## Focusing Instruction on Key Concepts Summary

Key concepts provide you with the opportunity to emphasize those curriculum outcomes and standards that have the greatest impact on student learning. After identifying key concepts, you can return to them throughout the year and monitor student learning as the year progresses. The key concepts approach offers a way to space instruction so that students constantly revisit concepts and, in doing so, add layers to their memory of it.

# CHAPTER 9

# Rethinking Year Plans

*Plan the year to interrupt the forgetting process.*

**DIGITAL SOLUTIONS**
Use an interactive whiteboard system to craft and post year plans. You can easily change the sequence and duration of concepts throughout the year.

In recent years, educators have realized that the year plan can be used to promote learning. Teachers are using them to structure their school year to ensure coverage of all curriculum outcomes and standards. Administrators are also using them to promote consistency from class to class. For both teachers and administrators, the important thing is that everything gets covered, which is a laudable goal.

Even so, ensuring that everything gets covered does not ensure that students learn everything. I would like to suggest that we can structure our school year in a way that actually helps students learn.

In common practice, outcomes and standards are divided into units of study. Within such a year plan, instruction is clustered into related concepts. Each cluster is assigned a continuous block of time during which the cluster is the focus of instruction. An example of this might be a focus on place value. All outcomes related to place value would together be the sole focus for four to six weeks. After that, the teacher might plan to focus on outcomes related to addition and subtraction for five to six weeks. Outcomes related to perimeter might get the next three-to-four-week block.

On the plus side, the units-of-study year plan allows the teacher to delve into concepts within a continuous block of time. Students can finish exploring these concepts before moving on to the next block. In addition, the units-of-study year plan works well with the structure of most textbooks.

**Addressing any concept just once does not bode well for retention.**

On the negative side, the units-of-study year plan permits concepts to be addressed only once during an entire school year. And addressing any concept just once does not bode well for retention.

There are many alternatives to the units-of-study year plan for mathematics. Here, I highlight two approaches that may be less familiar to educators but which may hold promise for boosting retention: the spiralled curriculum year plan and the key concept year plan.

## Strategy: Create a Spiralled Curriculum Year Plan

In a spiralled curriculum year plan, learning of particular concepts is spread out over time rather than being concentrated in shorter periods. Concepts are addressed several times throughout the year. During each revisit of the concept, the teacher and students explore the concept a little more deeply. Thus, learning moves from a shallow level at the beginning of the year to a deeper level by the end of the year.

By following a spiralled curriculum year plan, you *constantly review previously taught material* instead of always broaching new concepts in a sequential order. Constantly returning to concepts allows students to approach problems with confidence because, if students do not understand the concept the first time they learn about it, they have ample opportunity to engage with it more successfully later in the year. Students engage more competently with the concept with each visit.

When considering how to organize the year, we need not treat individual curriculum outcomes and standards in isolation. The spiralled curriculum year plan organizes teaching under either overarching questions or big ideas. An overarching question such as "How can I use numbers to compare and contrast?" serves as a way to group outcomes and standards. Big ideas are general mathematical ideas that support other ideas, highlight connections amongst mathematical ideas, or serve to illuminate mathematics as a human process (National Council of Teachers of Mathematics, 2000). An example of a big idea is "Numbers can be represented in a variety of ways." A small set of overarching questions or big ideas works well as the spine of a spiralled curriculum year plan.

Sometimes big ideas or overarching questions are preset by the jurisdiction within which a teacher works. In most cases, though, teachers are free to develop these on their own, by drawing from the themes and emphases within the curriculum.

Let's assume that you would like to develop a spiralled curriculum year plan organized under big ideas. Here are the main steps:

1. Craft between six and eight big ideas, depending on the grade level and curriculum.
2. Assign every outcome or standard to one big idea.
3. Divide the teaching time for each big idea into three sections.

To begin the year, you teach all the first sections for all your big ideas. Then you teach the second sections, and finally all the third sections.

When crafting a spiralled curriculum year plan, there are many decisions to be made. Below, I share the thought process I went through while crafting a spiralled curriculum year plan for an elementary class. This should help make explicit how to develop a plan that covers all concepts yet allows you to return to concepts through the year.

> By following a spiralled curriculum year plan, you constantly review previously taught material instead of always broaching new concepts in a sequential order.

## Designing a spiralled curriculum year plan

Looking through the grade level curriculum, there were many outcomes I had to consider. I wanted to group outcomes together in a way that would be meaningful for students, that would highlight the connections amongst the outcomes, and that would allow me to cycle through them throughout the year while increasing the complexity during each cycle.

I decided to group outcomes according to the big ideas in math. Each big idea served as an umbrella topic for multiple curriculum outcomes from multiple strands.

For each big idea, I examined the curriculum to identify the outcomes that fit well. While some outcomes could go under multiple big ideas, I decided to group the individual outcomes that had the strongest connections. But I made a point of looking for outcomes in multiple strands. Within each big idea, it was interesting to see various strands addressed. Through this process, I was making connections across the curriculum. Overall, I ended up with eight big ideas.

Next, I chunked up each big idea into three sections. The average time that I needed to spend on each section was approximately two weeks (with some sections requiring less time while others required a bit more, depending on the complexity of the idea and the learning needs of students). The key was to divide the teaching of each big idea into three sections, and cycle through the sections during the year.

The first section was more introductory, skimming the surface of concepts. After I cycled through the first section of all the big ideas, I then addressed the second sections of the big ideas. The second section went a bit more in-depth. After I addressed all the second sections, I then moved into the third sections. The third sections went deepest. It was like burrowing into the big ideas as the year progressed.

I was cognizant of not leaving students with the impression that they were just repeating concepts as they moved through each cycle. Instead, as students engaged with the same big idea three times during the year, they encountered different contexts each time. It was not about repeating the big ideas so much as it was about expanding the learning experiences of students throughout the year.

**By spacing learning, students must exert effort whenever they return to a concept and have to remember what they have already learned.**

Following a spiralled curriculum year plan enhances long-term retention. By spacing the learning, students must exert effort whenever they return to a concept and have to remember what they have already learned. In the typical units-of-study approach to planning, students encounter blocked practice. As we discussed earlier in the book, within blocked practice students may experience success in short-term performance, but this success does not indicate an ability to recall the learning in the future (Schmidt & Bjork, 1992). By spacing learning, teachers give students multiple opportunities to engage with a concept, thereby producing deeper, more conceptual understanding.

Another benefit of the spiralled curriculum approach is that students explore all concepts within each of the three learning cycles. With the year consisting of three cycles, or spirals, students will encounter concepts in combination with others at least three times. Students may, therefore, make deeper connections amongst the various concepts.

Take a moment and consider the following analogy. When grade level curriculum and standards are crafted, concepts are revisited from grade to grade. In each successive grade level, the expectations for the concept are deepened. Each grade level includes exploration of multiple concepts, and there exists a progression from one grade level to the next. This is spiralling curriculum across the grade levels. Spiralled curriculum *within* the grade level adopts exactly the same philosophy but with shorter cycles within one year.

The potential impact of a spiralled curriculum on student learning should not be underestimated. Learning occurs when curriculum outcomes and standards are connected, when one builds on another, and when they are revisited after a period of time.

Students will add layers of details to the memory of a concept during each visit. Because each concept is visited multiple times and multiple concepts are addressed in each cycle, instruction will be mixed, requiring students to take time to consider which concept is appropriate and how to apply the concept. Following a spiralled curriculum year plan supports all three stages of learning: encoding, consolidation, and retrieval.

**The potential impact of a spiralled curriculum on student learning should not be underestimated.**

---

### ELEMENTARY EXAMPLE: A TEACHER EXPLAINS

#### Designing a spiralled curriculum year plan

I wasn't pleased with my year plan. I had strategically placed each concept within the year so that there was a progression of learning and so that I would be able to teach all the curriculum. But—doesn't matter the year—I would have to leave out some outcomes as the year came to an end. Also, when I had to give a provincial test, I had to stop and make sure to re-address all the outcomes so that students would do well.

I just felt like I was constantly swimming upstream and that I was only covering curriculum instead of focusing on student learning.

So I needed to try something different. I decided to try a spiralled curriculum. I heard of it and thought that it would help.

Have to say, I am impressed. It was tough moving on to another concept without "finishing it" during the first few cycles because I wasn't used to moving on until most of the students fully understood the concept. But I was pleasantly surprised because, as the year unfolded, students kept demonstrating understanding whenever we returned to concepts.

During each cycle, students were remembering what they had learned earlier in the year and were able to apply this to the new learning. Also, students started to identify more and more connections between concepts because they kept revisiting them and were seeing concepts more frequently.

Finally, I didn't have to stop teaching to review before the provincial test. Students were already able to recall things, so I knew they were ready.

Not only was it great for time management in working with all the curriculum, it also felt as though my focus was on learning and not on covering curriculum.

I just needed to get over the fact that I could move on from concepts without having most students understand everything first.

---

## Strategy: Create a Key Concepts Year Plan

**By introducing key concepts at the beginning of the year, you and your students will have plenty of time to work with them.**

As discussed in Chapter 8, key concepts are curriculum outcomes and standards identified as being of the highest priority. Understanding of key concepts is the foundation on which understanding of other curriculum outcomes and standards can be built. Within a key concepts approach to year planning, key concepts are frontloaded in the year so that you and your students have plenty of time to work with them. Teaching of the remaining curriculum outcomes and standards follow later in the school year. If time allows, you can re-address key concepts near the end of the school year.

While a key concept approach to year planning bears some similarity to the units-of-study approach, there are considerable differences. First, all key concepts are taught first. They are most crucial to student success, so it makes sense that these concepts be introduced at the initial point of the year so that there is plenty of time for students to work with them and strengthen their understanding of them. Key concepts are strategically placed near the beginning of the year so that they can be revisited later in the year. It is this revisit that is the second difference to the units-of-study approach. Returning to key concepts helps students in recalling previous learning, adding layers to their memories, and strengthening the pathways to these memories.

By frontloading the year with key concepts, you can later use formative assessment data to decide which key concepts may need to be revisited. It also allows time for you to use that formative assessment to provide corrective feedback to students with misconceptions.

---

**INTERMEDIATE EXAMPLE: A TEACHER EXPLAINS**

### Transitioning to a key concepts year plan

It just makes so much sense. That is what I thought when I first heard about frontloading the year plan with key concepts.

Looking back, why would I wait until late winter or early spring to introduce a concept that I knew would help students with other concepts and that was critical to student success? If I must be honest, sometimes I would have to rush through these key concepts because I was behind in the year plan by the time they came around in the plan.

Now I spend the first part of the year on the key concepts so that students have plenty of time to work with them and I don't have to rush through them. Plus, I have found that doing these concepts early has helped students with other concepts that come later in the year.

The other thing I noticed was that I can return to some of these key concepts later in the year if students are struggling with them. My year plan now gives the students a lot more time to work with and learn the concepts.

When I see students are having difficulty with key concepts in the spring, I will give a few assessments and then provide corrective feedback. I wouldn't be able to do that with my old year plan because I didn't have the time to go back, as it was much later in the year when these concepts were introduced.

---

Consider the following key concept year plan discussion with a primary teacher, which offers insight regarding the thought process that goes into developing a key concept year plan.

## Developing a key concepts year plan

**The author:** Tell me about your key concept year plan.

**Primary teacher:** I wanted a year plan that prioritized key concepts. It would allow my students more time to develop their understanding of the concepts and would help them use this understanding to support them working with other concepts.

**The author:** How did you start the year?

**Primary teacher:** The first month was about numbers up to 1000. It was about having students get familiar with the numbers and how they compare. The focus was on representing through concrete objects, pictures, and numbers. Doing this, we also worked with estimation using referents. Overall, just understanding numbers that they will be working with.

After that, I moved into place value and understanding the sequence within numbers. Increasing and decreasing patterns worked well during this time as it tied in with place value and number sequence.

**The author:** How long would this take?

**Primary teacher:** About a month and a half. From there we would begin with addition and subtraction. I wanted to have students thinking about numbers, so instead of going into the standard algorithm like some wanted to, we reviewed addition and subtraction fact-learning strategies before moving into mentally adding and subtracting two-digit numbers. Was also good to include estimation as a way to predict sums and differences. From this, our work with addition and subtraction of numbers up to 1000 flowed well. Students were applying many of the mental strategies to solving these problems. We did this for about a month.

**The author:** Why did you choose that progression of learning?

**Primary teacher:** It flowed well, from representing and comparing numbers, to seeing this representation being applied to place value. And patterns in numbers helped so much when working with numbers up to 1000. My students became comfortable with numbers.

I also wanted to work with mental math strategies for addition and subtraction, to build these skills for students so that they could mentally solve problems.

**The author:** Makes sense.

**Primary teacher:** I then think that we have had a busy fall, so, after returning from winter break, we move into measurement. We talk about length and how to measure length using standard units, and then apply this to perimeter. Students really enjoy this hands-on time in math. This usually takes a month, because within January we also work in some ongoing review of the key concepts we addressed in the fall.

The next step is focusing on equations and talking about the meaning of equality. I find this works well before we spend about a month to a month and a half on multiplication and division. Within multiplication and division, we spend lots of time on using manipulatives and pictures to solve problems. We then represent this thinking in an equation.

Following March break, we work with fractions. This makes sense for the students as it connects well to division and fair sharing. Like multiplication and division, we use lots of concrete objects and pictures to represent our thinking. Takes about a month.

We continue to address lots of key concepts, working them into the year as it unfolds.

**The author:** Do students see the connections?

**Primary teacher**: Many do, and I highlight the connections as we move forward. We then focus on collecting data and making bar graphs. It's a great way to strengthen the discourse in the classroom, as up to this point we have explored lots of math concepts and can use this knowledge in our work. This works takes a few weeks.

**The author:** How long have you been approaching the year plan like this?

**Primary teacher**: For a few years. We conclude introducing new concepts with geometry. Lots of time spent exploring 3-D objects.

I do like to address the outcomes focused on time. I work these in throughout the year so that it is not a block of time so much as continuously referring to it and talking about it.

**The author:** Do you find that approaching the year in this way has given you more time with the key concepts?

**Primary teacher**: Definitely! After introducing them earlier in the year than I would have in other years, I have time to come back to them now and then. It helps me to determine how students are understanding the concepts and it helps them to keep seeing the key concepts.

**Analysis**:

Many primary teachers begin the year with patterns, partly because this is what many textbooks begin with. This teacher chose to start with representing and counting to 1000, which is a major key concept. He built on this base by addressing place value, another important key concept. Having provided a firm grounding, he was able to support students in making meaning when adding and subtracting instead of focusing on the standard algorithm.

Notice that the teacher did not choose to address multiplication and division or fractions in the first half of the year, as neither of these is identified as a key concept for this grade level. The teacher chose instead to focus on equality and inequality because it is a key concept—one that many teachers don't address until late in the year.

## Rethinking Year Plans Summary

Year plans are the structure for the school year. Too often, consideration of key concepts is confined within units of study. Such an approach compartmentalizes concepts and promotes a one-and-done perspective of learning. The two year-plan strategies I have shared—the spiralled curriculum and key concepts year plans—offer you the opportunity to prioritize significant concepts, revisit concepts, and use time effectively so that curriculum outcomes and standards are not omitted from the year. Both strategies have built-in spacing and mixing learning opportunities that will promote deeper mathematical understanding. Both strategies offer a way to organize the year so that your focus shifts from short-term performance toward long-term learning.

**CHAPTER 10**

# Rethinking Review

*Create spaced and mixed reviews to interrupt the forgetting process.*

**When to Use Mixed Review**
- during independent practice
- in small groups
- at centres
- for homework

Cramming does not have a stellar reputation: it might work for the test, but the memories don't last long.

Many people will say that they review concepts consistently in the classroom and that this review promotes student learning. Much of that review, however, can be characterized as focusing on short-term performance. Think about it. Before a quiz, exam, or standardized test, we spend considerable time reviewing concepts so that students will do well. Some might call that cramming. And cramming does not have a stellar reputation: it might work for the test, but the memories don't last long. How many times have you reviewed a topic with students, only to have them forget everything shortly after the test?

Another form of review is practice with concepts at the end of every unit. After spending time on a concept, you might provide students with questions that will require them to apply the concept. Unfortunately, this form of review is also ineffective. It is blocked practice (Rohrer, 2009), which does not require students to consider which strategy to apply, because students know that all the questions apply to the unit's concept. What this does is promote the illusion of mastery, as students develop a false idea of their understanding (Brown, Roediger III, & McDaniel, 2014). They do not have to pause and consider which strategy to apply to solve each question. The structure of most textbooks plays a role in encouraging this form of review. The review at the end of each unit examines only concepts addressed within that unit.

What is lacking in the traditional approach to review is spacing and mixing. As discussed in the Introduction, spacing means to spread concepts over time instead of clumping them together (Brown, Roediger III, & McDaniel, 2014). Mixing means to mix up questions so that consecutive questions cannot be solved using the same strategy (Brown, Roediger III, & McDaniel, 2014).

Mixed review places a greater cognitive demand on students, as they must think about the question, choose a strategy, and then apply it.

Consider what a mixed review would look like in the classroom. Students would return to concepts—possibly key concepts—multiple times over a period of time, so that they are constantly interrupting the process of forgetting. This mixed review places a greater cognitive demand on students, as they must think about the question, choose a strategy, and then apply it. This process strengthens students' memories by bringing new context to the original memory and strengthens the pathways to this memory because students must recall it many times. It may be that a student has not seen a concept for some time and will need to search their previous learnings to choose the best course of action.

The following comment from an intermediate teacher explains his approach to mixed review in mathematics.

---

**INTERMEDIATE EXAMPLE: A TEACHER EXPLAINS**

### Building a mixed approach to review

I was frustrated with students forgetting significant math concepts by the end of the year. Each year, I would have to do a week of review before the exam to ensure that students were well prepared and would remember the year's content.

The last few years, I have changed my approach. Now what I do is build opportunities for review throughout the year. I tried to include all concepts, but I narrowed it down to the big things that students need to know and that carry the most weight in the curriculum.

I have six or seven concepts that I include on each review, one or two questions for each concept, and I give it to my class every four to five weeks. What this does is students see the concepts throughout the year and must solve problems for each of the important ones.

In each review, I mix the order of the questions so that students don't get the same concept for several questions in a row. I want them to spend time thinking about the question, deciding what they need to do, and then doing it.

After each review, I go over the questions with the class. Overall, the mixed review takes 15 to 20 minutes every four to five weeks. Not a lot of time but a great value.

I must say that now, at the end of the year, I am not worried about going over everything that we worked with earlier. Instead, students have a better understanding of the concepts and are ready for the exam.

---

## Strategy: Create Daily Cumulative Reviews

**When to Use Daily Cumulative Reviews**
- before the lesson (minds-on activity)
- for homework

So how can we transform review into a vehicle for exposing students to a concept or strategy repeatedly throughout the year, instead of just a few isolated weeks during the school year?

One strategy that can accomplish this goal is a daily cumulative review. Steve Leinwand (2009) stressed the importance of giving students daily opportunities to review important mathematical skills and concepts. At the foundation of ongoing cumulative review is mixed review—multiple concepts are addressed, and questions for a concept are spaced across a period of time (Rohrer, 2009).

**DIGITAL SOLUTIONS**
You can create questions on a digital slide or in a word processing program, and then post them.

Mixed review has two central aspects: questions on a concept are spread across many review sessions, and each review session addresses a variety of concepts.

A form of daily cumulative review that I have found to be effective is a mixed review consisting of three to five somewhat brief questions that target multiple concepts (Leinwand, 2017). You can set this review to occur every day at the beginning of math class. Students would need about five minutes to solve the questions. You would need a further five minutes to review correct responses, either by demonstrating the process or by calling on students to share their thinking. If you check responses, you can provide students with feedback on which strategy they chose to solve each problem and how they applied the strategy.

By engaging students in daily cumulative review, you not only reinforce the material, you also give students a chance to clarify any misunderstandings they may have about the concepts. Further, this daily review enables you to collect ongoing data of students' progress. After watching trends in the data you collect, you can adjust instruction as necessary. Perhaps most important for retention, your students will be continuously recalling previous learning.

**ELEMENTARY EXAMPLE: A TEACHER EXPLAINS**

### Transitioning to a daily cumulative review

I spend a lot of time having to re-teach concepts at the end of the year. These concepts mostly include improper fractions and mixed numbers; area, perimeter, and volume; multiplication of two-digit by two-digit numbers; and place value of decimals.

So I decided to use daily cumulative review in my classroom. Each day, I dedicate the first ten minutes to it. The first five minutes are for students to solve the problems and the last five minutes are for me to share the solutions.

Some days I have students record their responses in their daily cumulative review notebook. When using their notebooks, students bring them up to me at the end of the 10 minutes and I make quick notes on my clipboard of what questions they didn't get correct. This helps me monitor progress. Other days, I have students use their whiteboards. When using their whiteboards, I ask students to hold the boards up and I quickly mark down which questions students answered incorrectly.

I was asked a lot at the beginning of the year if this was going to hinder my year plan. It doesn't. The first ten minutes were when we used to do a warm-up. Well, daily cumulative review has replaced that, and I couldn't be happier. In fact, I am working through the year plan very well, as I am not spending lots of time re-teaching these concepts throughout the year.

I was so pleased with how my students responded to this this last year. I looked through the standardized test results and saw that my students did very well with these concepts. They also did very well on questions where, although these concepts weren't the focus, they were still helpful for addressing the problem-solving situation.

I will keep going with this.

The following examples highlight how teachers applied daily cumulative review as a method to mix review and support students in recalling previous learning.

## Mixing concepts for a week of daily cumulative review

In this example, the teacher identified place value, perimeter, subtraction, and equations as the concepts she wanted to emphasize during the review. Notice how the concepts are arranged in a different order from day to day. (The teacher would continue focusing on these four concepts for several weeks, and then would revisit them occasionally during the rest of the schoolyear.)

| Monday | 1. What is the value of the underlined digit? 4<u>9</u>2 <br> 2. Find the perimeter of a square with a side length of 7 m. <br> 3. Solve: $568 - 197$ <br> 4. Find the value of ?      $13 = 6 + ?$ |
|---|---|
| Tuesday | 1. What is the value of the underlined digit? 80<u>9</u> <br> 2. Solve: $701 - 222$ <br> 3. Find the perimeter of a rectangle with a length of 5 m and width of 3 m. <br> 4. Find the value of ?      $12 = ? - 4$ |
| Wednesday | 1. Solve: $984 - 421$ <br> 2. Find the perimeter of a five-sided shape with each side length being 8 m. <br> 3. Find the value of ?      $? = 5 + 9$ <br> 4. What is the value of the underlined digit? <u>4</u>92 |
| Thursday | 1. Find the value of ?      $9 - 8 = ?$ <br> 2. Find the perimeter of a square with a side length of 8 cm. <br> 3. What is the value of the underlined digit? <u>6</u>39 <br> 4. Solve: $432 - 224$ |
| Friday | 1. Solve: $600 - 286$ <br> 2. Find the perimeter of a rectangle with a length of 5 cm and width of 8 cm. <br> 3. Find the value of ?      $? = 7 - 5$ <br> 4. What is the value of the underlined digit? 7<u>1</u>3 |

## Mixing concepts for a week of daily cumulative review

In this example, the teacher identified fractions, place value, decimals, and multiplication as the concepts she wanted to emphasize during the review. Notice how the concepts are arranged in a different order from day to day. (The teacher would continue focusing on these four concepts for several weeks and then revisit them occasionally thereafter.)

| | |
|---|---|
| Monday | 1. Write the following number in words: 583 839.34<br>2. Arrange the following in ascending order: 0.532, 0.523, 0.352, 0.325<br>3. Solve: $68 \times 34$<br>4. Place >, <, or = in the circle: $\frac{5}{6} \bigcirc \frac{7}{8}$ |
| Tuesday | 1. Arrange the following in descending order: 8.041, 8.140, 8.104, 8.014<br>2. Write the following number in words: 76 284.481<br>3. Place >, <, or = in the circle: $\frac{2}{5} \bigcirc \frac{2}{6}$<br>4. Solve: $42 \times 26$ |
| Wednesday | 1. Solve: $98 \times 98$<br>2. Place >, <, or = in the circle: $\frac{4}{7} \bigcirc \frac{5}{8}$<br>3. Arrange the following in ascending order: 7.22, 2.72, 2.27, 7.72<br>4. Write the following number in words: 900 290.075 |
| Thursday | 1. Arrange the following in descending order: 23.873, 23.837, 23.738, 23.783<br>2. Place >, <, or = in the circle: $\frac{3}{4} \bigcirc \frac{2}{3}$<br>3. Write the following number in words: 38.098<br>4. Solve: $56 \times 47$ |
| Friday | 1. Write the following number in words: 509.006<br>2. Solve: $38 \times 25$<br>3. Place >, <, or = in the circle: $\frac{3}{5} \bigcirc \frac{4}{9}$<br>4. Arrange the following in ascending order: 63.429, 63.294, 63.942, 63.924 |

## Mixing concepts for a week of daily cumulative review

In this example, the teacher identified order of operations, adding fractions, integers, and equations as the concepts he wanted to emphasize during the review. Notice how the concepts are arranged in a different order from day to day. (The teacher would continue focusing on these four concepts for several weeks and then revisit them occasionally thereafter.)

| Monday | 1. Arrange the numbers in ascending order: $\frac{5}{6}$, 0.532, 1 <br> 2. Solve: $\frac{2}{5} + \frac{1}{3}$ <br> 3. Solve: $5x + 8 = 18$ <br> 4. Solve: $(-3) + (-4)$ |
|---|---|
| Tuesday | 1. Solve: $(+3) - (+4)$ <br> 2. Solve: $3y - 2 = 13$ <br> 3. Solve: $\frac{5}{6} - \frac{1}{3}$ <br> 4. Arrange the numbers in descending order: $\frac{3}{4}$, $\frac{5}{6}$, 0.850 |
| Wednesday | 1. Solve: $7n + 3 = 24$ <br> 2. Solve: $(-8) - (+4)$ <br> 3. Arrange the numbers in ascending order: $\frac{8}{9}$, $\frac{7}{8}$, 0.900 <br> 4. Solve: $\frac{3}{4} + \frac{1}{8}$ |
| Thursday | 1. Solve: $\frac{6}{8} - \frac{1}{2}$ <br> 2. Arrange the numbers in descending order: 0.75, $\frac{5}{7}$, $\frac{4}{5}$ <br> 3. Solve: $(+9) + (-9)$ <br> 4. Solve: $2b - 9 = 17$ |
| Friday | 1. Solve: $4m + 3 = 19$ <br> 2. Solve: $\frac{4}{7} + \frac{1}{4}$ <br> 3. Arrange the numbers in ascending order: 0.535, $\frac{7}{8}$, $\frac{6}{7}$ <br> 4. Solve: $(+2) - (-2)$ |

## Rethinking Review Summary

To be effective, review needs to be spaced over time so that students return to concepts continuously and experience the desirable difficulty of retrieving this previous learning. Review needs to be mixed so that students can think about which concept is best suited to a particular problem and then apply it. A daily cumulative review accomplishes both of those goals. It can be applied successfully at any grade level. This small change in instructional approach can lead to significant improvement in student learning.

# 11

# Rethinking Testing

*Use frequent testing to spark more frequent retrieval.*

Traditionally, testing has been regarded as an assessment tool, as a measurement of student learning. Teachers measure student learning via tests at the mid-point of a unit, end of the unit, or the end of the term or year in the form of an exam. Tests, however, can be so much more than just a measure. They can also have a powerful impact on student learning (Roediger & Karpicke, 2006).

When students undertake a test, they are engaging in a form of retrieval. They must read the questions, search their memory for the previous learning, and then apply it to the task at hand. This test-taking retrieval practice helps strengthen memories as well as the pathways to these memories.

If testing is to be used as a strategy for improving retention, we have to adjust how we use them in the classroom. First, we can leave behind open-book tests. Closed-book tests provide students with much more desirable difficulty. When students are not permitted to use their book to search for answers, they must engage in a generative process by which they search their memory to recall the appropriate concept. Such a learning process would not occur had students been permitted to look up the answer in a book.

Second, we need to avoid the temptation to provide students with hints and prompts to help them do math questions. When we supply these, we remove some of the effort required in retrieving previous learning. So, while we may think of hints and prompts as helpful, what they actually do is hinder student learning because they remove the desirable difficulty that consolidates learning.

Third, we can ensure that tests are low stakes. If tests have a lot of weight associated with them, whether it be in terms of percentage of the final grade, rankings, or prizes, student anxiety will build. Anxiety erodes student ability to recall and apply previous learning. When you encourage students to view tests as low-stakes

**Tests can be so much more than just a measure. They can also have a powerful impact on student learning.**

learning tools, students are more apt to take risks, consider mistakes as learning opportunities, and use the experience to move them forward as learners.

Fourth, we can provide feedback, which plays a crucial role in student learning. Students need to be aware if their responses are correct or incorrect. This feedback can be generated by you, by students' peers, or through self-assessment strategies. Without such feedback, students might experience illusion of mastery. The timing for feedback is critical. It can be provided immediately, the same day, or within a few days but should not be so delayed that students have moved on from the learning experience and will not use the feedback to adjust their understanding.

---

**INTERMEDIATE EXAMPLE: A TEACHER EXPLAINS**

### Building a cumulative unit test

I always gave tests at the end of a unit. Didn't matter how long the unit was, students would be tested at the end. It would be worth a certain percentage of the course grade.

I always wondered if I was getting what I wanted out of the tests. If a student didn't do well on a test, it didn't change anything because we were onto the next unit. If a student did well on a test, I wasn't sure if they would be able to remember it in the future because we never went back to the concept.

So what I did was start using cumulative tests. I would identify key concepts for the year and then structure my tests to account for any of these that students had worked with. If the first unit was dealing with a key concept, that key concept would be the focus of the first test. Then, when we addressed the next key concept, the test for that unit would focus on the key concept for that unit but also have questions on the previous key concept. This would continue throughout the year so that each test addressing a key concept would have additional questions tagged onto it that targeted previously addressed key concepts.

This really helped students in remembering key concepts. They would receive feedback on how well they did and could use this information as they move forward.

---

The following sections will highlight two approaches to testing that move testing from an assessment tool to an instructional tool that supports students as learners. By making slight adjustments to how tests are used in the classroom, you can have a significant impact on student learning. The two approaches both allow for spacing and mixing of concepts as well as providing students with feedback that will help them build strong metacognition.

## Strategy: Design Cumulative Tests to Help Students Maintain Memories

**When to Use Cumulative Tests**
• during independent practice

Cumulative tests represent a shift in how tests are constructed and given to students. Primarily, they focus on the spaced and mixed testing of key concepts. The curriculum for each grade includes six to eight key concepts. After teaching one key concept you would provide students with a test addressing that key concept. You would do the same thing for the second key concept you teach. But you

**DIGITAL SOLUTIONS**

You can create tests using digital platforms that have forms or surveys. Using these, you can easily sort the questions and answers. This approach gives students immediate feedback.

would *also* address the first key concept. This process of tagging on previous key concepts to tests supports students in that the key concepts are being spaced out and mixed on tests. Students also receive feedback on their thinking in relation to all key concepts on every test.

The cumulative test approach focuses on key concepts for a few reasons. First, it is more manageable for you and your students to return repeatedly to a subset of concepts. Second, key concepts by definition support the learning of all curriculum outcomes and standards. Narrowing the focus on key concepts will support students in achieving all curriculum outcomes and standards.

|  | **Test 1** | **Test 2** | **Test 3** | **Test 4** | **Test 5** | **Test 6** |
|---|---|---|---|---|---|---|
| **Concepts Within the Test** | Key Concept 1 | Key Concept 2 | Key Concept 3 | Key Concept 4 | Key Concept 5 | Key Concept 6 |
|  |  | Key Concept 1 | Key Concept 2 | Key Concept 3 | Key Concept 4 | Key Concept 5 |
|  |  |  | Key Concept 1 | Key Concept 2 | Key Concept 3 | Key Concept 4 |
|  |  |  |  | Key Concept 1 | Key Concept 2 | Key Concept 3 |
|  |  |  |  |  | Key Concept 1 | Key Concept 2 |
|  |  |  |  |  |  | Key Concept 1 |

*Model for designing a series of cumulative tests*

The following graph displays common trends when applying a cumulative test approach to instruction. When students are tested on a concept initially, they perform well; when the concept is tested a second time, students do not do so well. This phenomenon stems from students being unused to being tested on a concept without having block practice immediately prior to the test. Over time, however, as students continue to encounter the concept on successive cumulative tests, their performance improves and they demonstrate greater understanding of the concept.

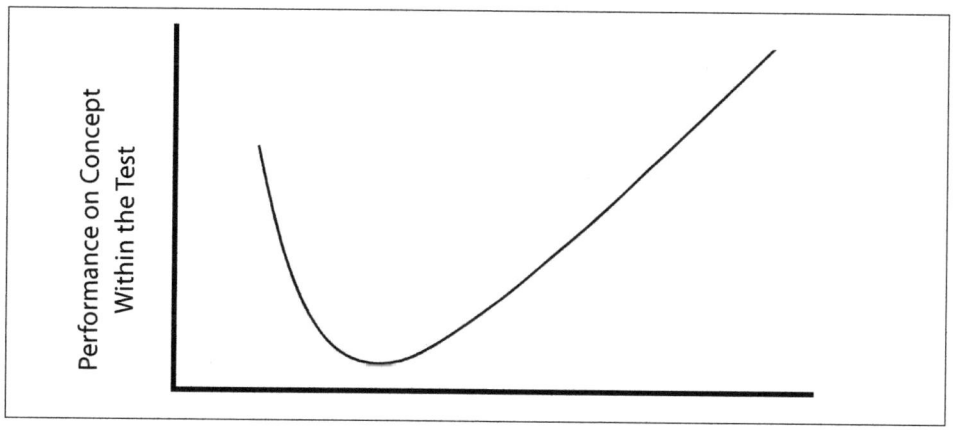

Number of Times Tested on the Concept

*Trends when using a cumulative test approach*

Consider the following cumulative test example provided by an elementary teacher.

---

**ELEMENTARY EXAMPLE: A TEACHER'S EXPERIENCE**

## Using a cumulative test approach

An elementary teacher had started the year by teaching the key concept of whole number place value. When she was finished, she provided students with the following questions, all related to key concept 1.

1. Write the following number in words:
   - 6473
   - 9002
2. Write the following numbers in expanded form:
   - 1901
   - 5740
3. Write the following numbers in standard form:
   - 6000 + 300 + 7
   - 10 + 200 + 8 + 7000
4. What is the value of the underlined digit?
   - 2693
   - 4107

In the following weeks, the elementary teacher then addressed the key concept of addition and subtraction. On the next cumulative test, you will notice that the most recent key concept is addressed first (key concept 2), followed by key concept 1.

1. Solve.
   - 2036 − 1963
   - 4276 + 1348
   - Manuel went to the sports centre and sold tickets for the school fundraiser. During the first week, Manuel sold 2390 tickets. He went back again for a second week and sold more tickets. Over the two weeks, Manuel sold 4953 tickets. How many tickets did Manuel sell in the second week?
2. Write the following number in words:
   - 7020
   - 4902
3. Write the following numbers in expanded form:
   - 3087
   - 5483
4. Write the following numbers in standard form:
   - 500 + 40 + 8000
   - 2 + 3000 + 600
5. What is the value of the underlined digit?
   - 2380
   - 9261

---

The elementary teacher then addressed the key concept of fractions. Here is the third test she provided students. The first two questions relate to the concept most recently taught: key concept 3. The rest follow the pattern of the previous test, addressing key concept 2 and finally key concept 1.

1. Arrange these fractions in ascending order:

   $\frac{3}{7}$    $\frac{4}{7}$    $\frac{2}{7}$    $\frac{5}{7}$

2. Arrange these fractions in descending order:

   $\frac{2}{9}$    $\frac{2}{3}$    $\frac{2}{4}$    $\frac{2}{5}$

3. Solve.
   - 6482 − 2683
   - 4236 + 2736
   - Alyah was given $5000 to purchase new items for the community park. If she spent $1700 on a swing set, $400 on a bench, and $2600 on a slide, how much money does she have remaining?

4. Write the following number in words:
   - 8937
   - 1920

5. Write the following numbers in expanded form:
   - 5301
   - 9623

6. Write the following numbers in standard form:
   - 50 + 1000 + 400 + 3
   - 7000 + 20 + 900

7. What is the value of the underlined digit?
   - 319<u>2</u>
   - <u>4</u>010

For the cumulative tests, the elementary teacher asked only a few questions for each key concept. Otherwise, the assessment would have been far too large for a student to complete.

## Strategy: Design Carry-Over Tests to Maintain Learning of Difficult Concepts

**When to Use Carry-Over Tests**
- during independent practice

How many times have you given a test and noticed that there were one or two questions that were not done well by most of your students? In response to such a situation, you might spend time going over those questions and modelling the correct application of the strategy. This may support students in the short-term, but what does it do for them in the long-term? While reviewing these questions and concepts immediately after the test is important, once is not enough. Given that students struggled with these concepts, they will need further opportunities to work with them.

Why not include the concept that students struggled with onto the next test? Beforehand, you could address this concept either through some practice, daily cumulative review, or exit tickets (to be discussed later, in Chapter 13). It makes sense to provide students opportunities to re-engage with the concept multiple times. Doing so provides the desirable difficulty needed to consolidate understanding.

By returning to problem concepts time and again, students have more opportunity to engage with them successfully.

By focusing further review and testing on the problem concepts, you provide students with further opportunities to engage with them. Each time, they will have to search their memory to recall the correct strategy and apply it to the question. Such an approach to testing involves spacing the concept out over time, mixing the concept with the concept being focused on within the test, and providing feedback to students that they can apply on the carry-over test. Testing thereby moves from being an assessment tool to being an instructional tool supporting student learning.

---

**INTERMEDIATE EXAMPLE: A TEACHER'S EXPERIENCE**

### Using the carry-over test strategy

An intermediate classroom was focusing on graphing and understanding the Cartesian plane (four quadrants). After checking the students' unit test results, the teacher noticed that many students struggled with the following question.

In which of the four quadrants do the following points lie?
- A (5, –3)
- B (6, 7)
- C (–4, 9)
- D (–1, –1)
- E (–6, 7)
- F (8, 6)
- G (–5, –7)
- H (3, –6)

The teacher applied the carry-over test approach to focus students' minds on the Cartesian plane concept, which so many students had struggled with. Although the focus of the next unit would be operations including decimals, the teacher tacked the following questions onto that unit's test.

Identify if the following statements are true or false.
- The coordinate (–4, 2) lies in quadrant 3.
- The coordinate (5, 1) lies in quadrant 1.
- The coordinate (–3, –6) lies in quadrant 4.
- The coordinate (7, –5) lies in quadrant 2.

When the test was checked, the teacher noticed that considerably fewer students struggling with the Cartesian plane question.

Depending on the results, the teacher might decide to add further review or add another Cartesian plane question onto the next unit test to ensure students have maintained this learning. The teacher would only add one or two questions addressing previous concepts to successive tests so as not to overwhelm students during the testing process.

---

## Rethinking Testing Summary

Tests play a significant role in education. We can expand that role from merely assessment tool to instructional tool. Cumulative tests and carry-over tests both provide students with an opportunity to recall previous learning and to apply it to new situations. By repeatedly accessing concepts learned earlier in the year, students refine their understanding of these concepts and strengthen their ability to recall them. Either of these testing approaches will help students focus on long-term understanding as opposed to short-term performance.

# CHAPTER 12

# Rethinking Homework

*Reimagine homework as an opportunity for review.*

Homework is a simple concept: an out-of-classroom activity intended to provide students with an opportunity to practice concepts they are learning in school. For such a simple idea, homework can be a contentious topic, exciting a lot of passion. Some schools assign homework while other schools do not. Of the schools that do choose to assign homework, primary teachers may choose not to assign homework while elementary teachers choose to do so. What tends to be the deciding factor is whether the teacher believes students can work on concepts independently.

When researching the impact of homework on student learning, Hattie (2009) determined that the effects of homework are twice as large for secondary students as they are for intermediate students, and twice as large again for intermediate students as they are for elementary students. Many educators concluded that the less homework for elementary students, the better.

Before you consider whether to assign homework, let's consider its purposes and possible benefits. Here are a few possible purposes:

**Teachers should always consider the purpose of the homework they are considering assigning.**

- **to prepare**: students gather information or materials for future learning activities in the classroom
- **to practice**: students practice concepts recently explored in the classroom
- **to enrich**: students apply a concept in novel situations

Teachers should always consider which purpose is underlying the homework they are considering assigning. Perhaps the most valuable purpose is to enrich, because it will help students retain their memories of concepts that they have already worked hard to learn.

## Strategy: Assign Homework That Supports Recall of Previous Learnings

Many teachers assign homework nightly with the expectation that students will complete the assignment before the next day of class. In such a situation, homework assignments tend to be brief. Another approach would be to provide tasks for students at the beginning of the week with a deadline of the end of the week. Within a week's timeframe, the homework can be more complex and time-consuming, as students have more time to work on the task. They can return to the task if they hit a stumbling block and need assistance or simply a break from the work.

The structure of the assigned task can help you make use of homework as a learning strategy. Steve Leinwand (2017) provided a template for homework that involves a range of tasks.

**DIGITAL SOLUTIONS**
Use an interactive whiteboard system to house homework questions. Students can post their replies.

- First are two problems requiring a new skill. These two problems are enough for students to approach the concept while avoiding the possibility of reinforcing a misconception.
- Second come four problems requiring recall of concepts learned in previous lessons. These four problems serve the purpose of mixing the review, whereby students need to retrieve previous learning from memory.
- Third, students are provided two problems requiring extra work or explanation. These two problems serve the purpose of affording students the opportunity to problem solve.

Another structural option for the homework plan is a cumulative review approach. You would provide students with three to four problems that focus on three to four concepts you have identified as being crucial to student mathematical understanding (that is, key concepts). You could choose key concepts explored earlier in the year and concentrate all homework on these key concepts. You could focus on the same key concepts throughout the entire year, or you could cycle through various key concepts during the year using bi-weekly or monthly blocks.

Whichever homework option you choose, it is important to provide multiple opportunities to *check* homework. The act of checking homework validates it and provides students feedback on their responses, which in turn supports accuracy in metacognition. Regardless of the timeframe or structure of the homework plan, all approaches provide you with opportunities for formative assessment. You can check the homework together as a class, collect homework assignments for marking, or provide students with opportunities to self-correct. The data you collect can help you plan whole-class or small-group additional instruction.

Homework can have a big impact on student learning. When concepts are spaced and mixed throughout the year, students are provided opportunities to recall previous learning and apply it to solve questions. The more they engage with such a mixed review of previously learned concepts, the stronger the learning becomes and the more the pathways to this learning are strengthened.

Keep in mind that students' home situations may vary widely. Some students in your classroom may live in situations that are not conducive to completing homework. If you are aware of difficulties, avoid exacerbating inequalities by making accommodations to support students in completing homework at school.

## Focusing nightly homework on key concepts

Homework was always a bone of contention for me. I knew that there were benefits to it, but I was not seeing any of those benefits in the classroom.

So, a few years ago, I completely revised my approach to homework. Instead of assigning nightly questions from the book that covered the topic we had explored that day, I started to assign my homework on Monday and required students to complete it by Friday. It asked questions about concepts learned earlier in the year that I knew students would need to know well to have success in the grade. I shared that we would check the responses on Friday, and that it wouldn't be for marks.

I went through my planning guide and identified eight important concepts. What I did was address five of those concepts every week with five questions. It was five of the eight concepts one week, and a different five of the eight concepts the next week.

What I noticed, over time, was that students were getting really good with these eight concepts, which helped them apply those concepts throughout the year.

The following examples highlight how teachers applied cumulative review as a method to mix review and support students in recalling previous learning.

## Planning a weekly homework assignment

The teacher assigned the homework on Monday and required students to have it completed by Friday. The homework always consisted of two parts: Part 1 addressed key concepts covered earlier in the year, and Part 2 addressed concepts that were currently being explored. Such an approach to homework reflected a spaced and mixed approach to instruction and learning. On Friday morning, the teacher would check the homework during the first part of the math lesson.

**Part 1**
- Write the value of each digit in 781.
- Draw base-ten blocks to show 563 in two different ways. You can use a square for a flat, a stick for a rod, and a dot for a unit.

**Part 2**
- Solve: $674 + 193$
- Solve: $301 - 137$
- Solve: $902 - 683$
- Solve: $339 + 428$

## Planning a nightly homework assignment

A teacher assigned four problems for homework nightly. The problems dealt with concepts that had been addressed earlier in the year. This strategy was an attempt by the teacher to continuously space and mix concepts to support student learning. She began every math class by posting the solutions to the problems on the board. Students would check their own work. This checking of homework took only about five minutes.

- Solve: 64 x 52
- Solve: 452 ÷ 6
- Place >, <, or = in the circle: $\frac{3}{5}$ ◯ $\frac{3}{7}$.
- Write the value of each digit in 18.098.

## Planning a nightly homework assignment

The teacher assigned homework nightly. The first part of the homework addressed the concept being explored in class currently. A second part addressed concepts from earlier in the year. Finally, a third part consisted of problems requiring a little more concentrated effort. The teacher designed this plan to provide mixed and spaced retrieval opportunities that would make learning effortful. The teacher posted solutions on the classroom whiteboard the next morning.

**Part 1**
- Solve: $4b + 7 = 35$
- Solve: $8a - 3 = 53$

**Part 2**
- Solve: $(-2) + (-5)$
- Solve: $(+6) + (-4)$
- Solve: $(-3) - (-7)$
- Solve: $(+8) - (+9)$

**Part 3**
- Create a table of values for the expression $3y + 2$.
- Draw a graph for the above expression.

## Rethinking Homework Summary

If you decide to assign homework, have a clear purpose in mind and give the homework a structure that supports students in meeting this purpose. You can mix and space review of key concepts on a nightly or weekly rotation. In this way, homework becomes an instructional strategy to consolidate and retain concepts not only from current lessons but also from previous lessons.

# 13

# Adjusting Common Classroom Tools and Strategies

*Apply a retrieval practice approach to common instructional tools and strategies.*

**Tweaking even a few aspects of your instructional approach can give it a much larger impact.**

Virtually all classroom tools and strategies can be remade as vehicles for supporting students as they encode, consolidate, and retrieve learning. In this chapter, we will examine several common instructional tools and strategies and explore how they can be used to boost the retrieval capacity of your students. Tweaking even a few aspects of your instructional approach can give it a much larger impact.

To begin, you might consider simply spacing out the teaching of a concept over time, interspersing it with your teaching of other concepts. The difference may seem small, but the potential value is great. Please note that any change you make will take time to produce the results you want. In fact, the initial performance of students may disappoint you, because students may be perplexed and skeptical. If you suddenly begin giving them homework questions unrelated to the unit they are currently studying in class, for example, they may even feel a little betrayed to get these "surprise" questions. Over time, however, students will demonstrate understanding as they improve their ability to recall with every retrieval practice opportunity.

### Getting over students' initial wariness

I had heard about supporting children in being able to recall previous learning and I wanted to do this in my classroom. However, I was overwhelmed at making a change that I thought would take a lot of preparation and time to do. I was wrong. It wasn't anything big, just more strategic in how I approached my teaching.

I used exit tickets a lot in my classroom. So what I did was start asking students questions about concepts they had encountered earlier in the year instead of that day. The new approach surprised students at first, and they didn't do that well. But over a few weeks (I gave one exit ticket a week), students started to do very well recalling things.

I now use the strategy in my morning message, where I bring up a math concept that they haven't seen in a while, and it is going great.

It really is a small change but makes such a big, big difference.

## Strategy: Using Exit Tickets to Spark Recall

**When to Use Exit Tickets**
- after the lesson (consolidation)

Exit tickets are a formative assessment tool we typically use to prompt students to demonstrate their understanding of a lesson. Exit tickets usually consist of a question or two designed to give a teacher a quick gauge of student learning. These tools do provide teachers with an inkling of the degree to which students understand the day's lesson. These in-the-moment snapshots of student learning, however, cannot predict or ensure long-term performance.

So why not switch things up? You can easily turn the exit ticket into a vehicle for spaced and mixed review. Instead of asking students to recall something about the day's lesson, ask them to recall something they learned the previous week or month! You would be providing students with opportunities to experience desirable difficulties that would strengthen their learning.

Here are a few models of prompts you can use to develop exit tickets for your primary or intermediate students.

**DIGITAL SOLUTIONS**

An interactive whiteboard system can be used so that students can post their responses. This approach allows you to identify trends within the replies.

- How is the concept we explored today similar to another concept you have learned?
- How is the concept we explored today different from another concept you have learned?
- Is there another concept that you could have used to solve problems like the ones we explored today? How do you know?
- What can you tell me about _____ [concept previously explored]?
- [After providing a problem to solve.] Solve this problem using _____ [concept previously explored].
- [After providing students with a work sample of a previously explored concept that includes errors] Identify which parts of this answer are done correctly and which parts are done incorrectly.

# Strategy: Teaching with Then-Now-Later Grids

**When to Use Then-Now-Later Grids**

- during the lesson (independent practice)
- for homework

Then-now-later grids are designed to help students recall previous learning (then), address current concepts (now), and begin to explore concepts identified for the near future (later). These grids provide three questions in all three categories, for a total of nine questions. Students are expected to complete the grid within a set period of time, usually a week.

By providing students with challenges in all three categories at the same time, students' exposure to concepts is mixed because not all questions and prompts focus on the same concept. Further, concepts can be cycled through the three categories over consecutive weeks as the concepts transform from future to present to previous learning, so students' exposure to concepts is also spaced.

For a few reasons, then-now-later grids should probably be established for a week in duration. First, you don't have time to be changing up the questions more than once a week. Second, a weekly cycle provides students flexibility in when to address the questions and prompts—they may choose to work on one category for some time and then turn to the others. Third, when not rushed, students have time to dig more deeply into questions, a luxury that might not be possible if the cycle was a day or two. The grid itself is usually a 3 × 3 square with three rows and three columns, so each square has three questions for each category for a total of nine questions.

| Then | Then | Then |
|------|------|------|
| Now | Now | Now |
| Later | Later | Later |

*Structure for a then-now-later grid*

Simply by arranging a variety of questions and prompts—which you probably already have—into a then-now-later grid, you require students to identify an appropriate strategy to solve each problem, retrieve it from their memory, and then apply it. Plus, the "later" category will have your students exploring their memory for previous learning of similar concepts that will support them in reaching a solution. This challenge—to solve a problem without first being taught the strategy—is one of the learning strategies discussed in Part 1.

The following are examples of the then-now-later grid applied at different grade levels. Each of these examples highlights how the teacher approached instruction from a spacing and mixing perspective.

**DIGITAL SOLUTIONS**

A digital version of a now then-later grid can be tailored easily to suit the needs of individual students.

## Designing a then-now-later grid for primary students

A primary teacher was covering addition and subtraction. Previously, she had taught a unit on patterns. Coming up next in her year plan was multiplication. The following is one of the then-now-later grids she used during that time.

| | | |
|---|---|---|
| Fill in the blank: 326, 336, 346, ___, ___, ___ | Fill in the blank: 814, ___, 794, 784, ___, ___ | Fix the errors in the following: 557, 562, 568, 572, 577, 584 |
| Find the sum of 273 and 493. Find the difference between 287 and 890. | Find the sum of 579 and 198. Find the difference between 545 and 700. | Solve: 274 + 498 Solve: 850 − 198 |
| On the football field, there are four equal groups of 8 students. How many students are on the football field? | There were 7 rows of tulips planted in the garden. If each row has 9 tulips, how many tulips are in the garden? | Write a multiplication sentence for the following visual:  *  *  *  *  *  *   *  *  *  *  *  *   *  *  *  *  *  * |

## Designing a then-now-later grid for elementary students

An elementary teacher had worked on place value of whole numbers and decimals earlier in the year and was now focusing instruction on mixed numbers and improper fractions. Coming up next in the year plan was creating tables of values. The following is an example of one of the then-now-later grids he used during that time.

| | | |
|---|---|---|
| Write the following number in words: 67 585.903 | Record the value of each digit in the following number: 582 197.064 | Using base-ten blocks, draw a picture of the following number: 0.965 |

| Express $\frac{13}{5}$ as a mixed number. | Express $6\frac{3}{4}$ as an improper fraction. | Express $\frac{29}{7}$ as a mixed number. |
|---|---|---|
| Complete a table of values for the expression 4x. | Complete a table of values for the expression 6a − 4. | Complete a table of values for the expression (c + 2) ÷ 2. |

### Designing a then-now-later grid for intermediate students

An intermediate teacher had worked on integers earlier in the year and was now focusing instruction on addition and subtraction of fractions. Coming up next in the year plan was multiplication of whole numbers and decimal numbers. The following is an example of one of the then-now-later grids she used during that time.

| Solve: (−9) + (−8) | Solve: (+1) + (−5) | Solve: (−4) − (−7) |
|---|---|---|
| Solve: $\frac{5}{6} - \frac{2}{3}$ | Solve: $\frac{2}{7} + \frac{1}{3}$ | Solve: $\frac{5}{8} - \frac{1}{2}$ |
| $5632 \times 1 = 5632$ How does knowing the above equation help solve $5632 \times 0.1$? | $64 \times 0.1 = 6.4$ How does knowing the above equation help solve $64 \times 0.001$? | $7462 \times 1 = 7462$ How does knowing the above equation help solve $7462 \times 0.001$? |

## Strategy: Posting Choice Boards

**When to Use Choice Boards**
- during the lesson (independent practice)
- for homework

A choice board is a visual presentation of multiple questions or prompts that address one or more concepts. It gets its name because students are typically allowed to choose which questions or prompts to solve. Some choice boards have guidelines: students must select a certain number of questions or prompts from a particular row or of a particular color.

To create a choice board that addresses a learning perspective that includes encoding, consolidation, and retrieval, create questions for the board that cover a variety of concepts that students have already explored. Arrange the questions on the choice board in such a way that students must select questions from a variety of concepts. (For example, you might put a different concept in each row and then require students to select at least one question from each row.) By doing this, students will create their own mixed review, thereby increasing the desirable difficulty of the exercise, which leads to stronger learning.

By providing multiple consecutive choice boards throughout the term or year, students' exposure to concepts will be both spaced and mixed—two integral components of retrieval practice.

Note that choice boards can be posted as a large visual in the classroom, where all students can work with it, but they could also be distributed on paper (a good option if used for homework). Note that a paper or digital version lends itself to tailoring, so that each student has a choice board with questions targeting their unique learning needs.

The following choice board was used in an elementary classroom. The teacher asked students to complete all questions in one box from every row. Students could also be asked to complete all the questions in one box from each column, or in all three boxes in a row, column, or diagonal of their choosing.

As fractions is a major concept, the teacher spread out instruction on fractions over a long period of time using a spiralled curriculum. The use of a choice board was a way for the teacher to mix the applications of fractions and to have the concept spaced over time.

## Elementary Choice Board
### Working with Fractions

| Represent | Label | Compare |
|---|---|---|
| Draw a picture to represent the following fractions:<br>a. $\frac{3}{4}$<br>b. $\frac{1}{3}$<br>c. $\frac{7}{10}$<br>d. $\frac{5}{8}$ | Write a fraction for the following picture:<br>a.<br>b.<br>c. | Place $>$, $<$ or $=$ between the following:<br>a. $\frac{1}{4}$ $\frac{1}{5}$<br>b. $\frac{1}{3}$ $\frac{3}{7}$<br>c. $\frac{4}{5}$ $\frac{5}{6}$<br>d. $\frac{6}{8}$ $\frac{3}{4}$ |
| **Order**<br>Arrange the following fractions in descending order:<br>a. $\frac{1}{5}$ $\frac{4}{5}$ $\frac{3}{7}$<br>b. $\frac{3}{6}$ $\frac{5}{8}$ $\frac{2}{3}$<br>c. $\frac{3}{6}$ $\frac{3}{5}$ $\frac{3}{8}$ | **Number Line**<br>Place the following fractions on a number line:<br>a. $\frac{1}{5}$ $\frac{4}{5}$ $\frac{2}{5}$<br>b. $\frac{11}{5}$ $\frac{9}{6}$ $\frac{1}{3}$<br>c. $\frac{3}{4}$ $\frac{9}{12}$ $\frac{13}{7}$ | **Fractions & Decimals**<br>Express the following fractions as decimals:<br>a. $\frac{3}{5}$<br>b. $\frac{1}{4}$<br>c. $\frac{9}{10}$<br>d. $\frac{3}{4}$ |
| **Fractions & Mixed Numbers**<br>Express the following improper fractions as mixed numbers:<br>a. $\frac{13}{3}$<br>b. $\frac{21}{6}$<br>c. $\frac{9}{4}$<br>d. $\frac{20}{7}$ | **Equivalent Fractions**<br>Write 2 equivalent fractions for each of the following:<br>a. $\frac{5}{6}$<br>b. $\frac{1}{4}$<br>c. $\frac{6}{10}$<br>d. $\frac{2}{4}$ | **Mixed Numbers & Fractions**<br>Express the following mixed numbers as improper fractions:<br>a. $2\frac{1}{3}$<br>b. $1\frac{2}{5}$<br>c. $3\frac{4}{7}$<br>d. $4\frac{2}{5}$ |

*Model of an elementary choice board*

# Strategy: Use Think-Pair-Share for Review

**When to Use Think-Pair-Share**
- during the lesson (independent practice)
- after the lesson (consolidation)

To launch a think-pair-share activity that would encourage consolidation and retrieval, all you need to do is provide students with a concept they learned earlier in the year or in a previous grade.

Many teachers use the think-pair-share instructional strategy in their classrooms, to engage students in reading, writing, and math alike. This strategy provides students with the opportunity to think on their own, then collaborate and share their thinking with a partner or small group for the purpose of refinement, and then a sharing of this refined thinking with the whole class.

To launch a think-pair-share activity that would encourage consolidation and retrieval, all you would need to do is provide students with a concept from earlier in the year or previous grade. Students think about the concept, remembering as much as they can about it; share their memories with a partner or small group; and then share again with the whole class.

The initial prompt you choose could simply be the name of a concept. Alternatively, you could make it more open ended or more restricted. (For example, you could focus on comparisons, connections, and reflections.) By holding think-pair-shares about various previously learned concepts over time, you would ensure that students recall information while strengthening their memories of that concept at the same time.

The following is an example a think-pair-share activity from an intermediate grade. This activity is a great support for spacing and mixing—spacing, in that you can return to concepts throughout the year; and, mixing, in that you can interject different concepts in between your focus on other concepts throughout the year.

---

**INTERMEDIATE EXAMPLE: A TEACHER'S EXPERIENCE**

### Using think-pair-share to retrieve memory

An intermediate teacher wanted to provide an opportunity for students to recall learning related to surface area. This was a concept that students had not engaged with in quite some time. The teacher chose to have students undertake a think-pair-share.

- **Think**—Students were first asked to consider how they would approach a problem that required them to find the surface area of a three-dimensional object. The directions stipulated that students could not use any resources to assist them in recalling this previous learning. The teacher directed students to record their thinking on paper.
- **Pair**—Students were then assigned a partner with whom they could share their thinking about the prompt. Both students read each other what they wrote. They were encouraged to discuss their ideas, change their minds, rationalize their thinking, and ask each other questions to refine their thinking of surface area. Crucially, partners were encouraged to come to an agreement about how they would find the surface area of a three-dimensional object.
- **Share**—Partners shared their agreed-upon thinking with the class. This was another opportunity for students to listen to the thinking of others and to refine their approach to surface area if necessary.

---

The strength of this think-pair-share activity is that students were required to recall previous learning of surface area from memory. This effortful learning activity would help students to re-enter their memory of surface area and to enhance the pathways to that learning. When discussing surface area with a partner, and later with the class, students would hear others' understandings of surface area, which would assist them in retrieving further memories of the concept or adding new layers of detail to their understanding.

## Strategy: Alternating Figure-It-Out and Here-It-Is

**When to Use Figure-It-Out and Here-It-Is**

- during the lesson (independent practice)
- after the lesson (consolidation)

By offering students both figure-it-out and here-it-is, you ensure that they experience the type of problem from more than one angle.

On the one hand, we know that there is much to be gained in having students figure out a problem for themselves. They explore possible strategies, make connections, reflect on learning, and strengthen their metacognition during such tasks. On the other hand, we also know that there is much to be said for walking students through a completed work sample or modelled solution so that they see the steps involved and have an opportunity to question the process.

The figure-it-out and here-it-is strategy is simply a matter of alternating your use of these two scenarios. To begin, you might provide a question and have students select a concept or strategy and apply it themselves. Next, you might work through a similar problem, doing either a think aloud or a deconstruction of a work sample. By offering students both scenarios, you ensure that they experience the type of problem from more than one angle. They strengthen their understanding of the concept as they alternately work through it themselves and then watch another person do so.

Figure-it-out and here-it-is can be interchanged in a single class, during a week, or throughout the term or year, so that the students engage differently with the concept from time to time. The mix of desirable difficulty generated by the figure-it-out approach nicely pairs with the consolidation supplied by the here-it-is approach. (If students used an unfamiliar strategy to reach the right answer, they would see that there is more than one way to reach the same answer.)

The following is an example of the figure-it-out and here-it-is strategy applied in a primary classroom.

### PRIMARY EXAMPLE: A TEACHER'S EXPERIENCE

#### Applying figure-it-out and here-it-is to encourage facility with word problems

A teacher wanted to reacquaint students with word problems, which they had covered earlier in the year. So he decided to facilitate a two-day activity on word problems. On Day 1, the teacher provided students with the following word problem and asked them to solve it on their own.

- A group of students wanted to collect books for the school library. They set a goal of 800 books to collect. The 13 students collected 342 books in the first month and another 379 books in the second month. How many more books do the students have to collect to reach their goal?

On Day 2, the teacher provided a similar word problem and shared her approach to solving the problem through a think aloud.

- There were 673 fans in attendance at the local hockey game. Of these 673 fans, 598 had purchased a ticket and the remaining fans had won their ticket. Of the fans that bought tickets, only 276 were cheering the home team. How many of the remaining fans who bought a ticket were cheering for the visiting team?

By varying the two approaches used during instruction, generative on Day 1 and modelled on Day 2, students engaged in two different learning models. By alternating the instruction, the teacher was making learning effortful. Students were not just following an instructional routine. Instead, there were different expectations each day, thus adding to the amount of effort students needed to apply to recall previous learning before using it to find a solution.

## Adjusting Common Classroom Tools and Strategies Summary

Each of the examples highlighted within this chapter can help you frame instruction that supports your students with encoding, consolidating, and retrieving previous learning. With relatively minor adjustments in how you handle things in the classroom, you ensure that students approach math questions and problems in a variety of ways. By doing so, you help students to refine their understanding of concepts and strengthen the pathways to this understanding.

# Part 2 Conclusion

**To support students pursuing mathematical understanding, the instructional strategies you employ must be sound and purposeful.**

You play a significant role in student learning. In Part 2 you encountered instructional strategies you can use to support your students in all three stages of learning: encoding, consolidation, and retrieval. You can use them to introduce spaced and mixed practice into your classroom. These strategies can help your students add details to their previous learning, refine their previous learning, retrieve previous learning from memory, and apply this learning to solve novel problems. In other words, your students will build their long-term retention.

None of the instructional strategies presented in Part 2 need work in isolation. On their own, any of these strategies will support you in creating a classroom in which students learn for the long-term. Using them in combination, however, gives you a better chance of increasing the effectiveness of your instruction and reaching all your students.

The strategies can and do work well together. For example, by establishing key concepts, you can craft a year plan that is frontloaded or spiralled; structure review that focuses on the significant curriculum outcomes and standards; and create testing opportunities that support learning and metacognition. Similarly, by shifting your assessment focus to formative assessment through conversation, observation, and review of product, you can glean a more comprehensive understanding of student learning. By better knowing your class as a whole and your students individually, you can better choose from a variety of strategies—such as exit tickets, think-pair-share, and figure-it-out and here-it-is—to tailor your instruction to your students' specific needs.

The first step you take toward strengthening your instructional decisions can be whatever size you wish. You might begin by providing students with weekly then-now-later grids that address concepts they encountered earlier in the year. Or you might craft a homework plan that includes a mixed review approach to practice. Regardless, once you begin this journey, you will see its impact and will continue to notice subtle changes in student retention. As time goes on and you incorporate more of these strategies into your classroom routines, you will find the instructional strategies that work best for you and your students.

# Putting It All Together

It is time that we increase the value of our instruction, so that we can see the gains in student learning that we seek.

I am in awe of the time and effort teachers invest in their students. Within the first few paragraphs of the introduction to this book, I had referenced how hard teachers work to meet the learning needs of their students. I don't like to see teachers' extraordinary efforts have less impact than they should have. To increase our effectiveness, we need to work smarter.

We have many instructional and learning tools at our disposal. So how can we apply these tools to improve the effectiveness of our teaching? In this book, I have presented learning and instructional strategies that can help you work smarter in the classroom. My goal has been to identify and highlight options for the classroom that are manageable and sustainable. It is time that we increase the value of our instruction, so that we can see the gains in student learning that we seek.

Many classrooms are focused on short-term performance—such as answering questions correctly or doing well on a test—instead of learning. While short-term performance feels great, the material has not actually been learned if students are not retaining the information and applying it to future problems. Learning, as defined in this book, has focused on students encoding information, consolidating it in their minds (either as new learning or layers added to previous learning), and then retrieving and applying it in the future. Helping students learn something thoroughly requires more effort than working towards short-term performance, but it may be the "answer" that teachers are seeking.

Consider how this broadened definition of learning would change how you handle review in your classroom. Students would work with concepts repeatedly over a duration of time (spaced review), and they would address multiple concepts in each review (mixed review). Your students would have to understand a problem, decide on an effective strategy, and then apply their chosen strategy. Such an approach to instruction and learning requires more effort by the student, thereby supporting their development as independent learners. Within this model of instruction and learning, students strengthen their understanding and their ability to recall their learnings to help them solve future problems.

## Matching Our Plans to the Goal

What we have is an instructional issue. Recognizing this, it is incumbent on us, as teachers, to examine our instruction and make the necessary changes.

If our instructional goal is to enable our students to be independent learners who can apply concepts from memory, then we need to adjust our instruction to help us achieve that goal. I have learned from many teachers that they continually hear students say they forget a concept learned earlier in the year or previous grade level and cannot possibly apply it to solve a problem. If we heard this from a few students now and then, we could hypothesize that we have an individual learning issue. But we're hearing and seeing this everywhere, within classes, across grades, and among multiple schools. So what we have is an instructional issue. Recognizing this, it is incumbent on us, as teachers, to examine our instruction and make the necessary changes.

To achieve our goal to turn students into independent learners able to apply concepts from memory, we need to introduce effort. By spacing and mixing concepts throughout the year, students will experience desirable difficulties, which help them learn. A natural part of the learning process is forgetting and then remembering again. We can generate those opportunities to interrupt the forgetting process. By doing this, we will support students in making math *stick*.

## From the Classroom

The following comments were provided by teachers who have brought into their classrooms the learning approach outlined in this book. They tell of the remarkable changes they witnessed.

**PRIMARY EXAMPLE: A TEACHER EXPLAINS**

### Success focusing on review and formative assessment

What I was doing wasn't working. It was time for a change. So I made a change to work with key concepts. Before the year began, I looked through the grade level content and identified eight key concepts that students would need to master for them to have success this year and next.

I spent more time with these concepts as they came up during the year. Sometimes I would throw a few related questions on the SMART Board® to see if students were remembering how to apply these key concepts.

To go along with this, every now and then I would ask students to write what they knew about the key concepts. At first, students found it hard to remember, but over time their writing had more and more details. They were getting better with the eight concepts as the year unfolded.

As part of the writing activity, I would have students draw pictures to help them understand the concept better. I'd just give them a math term associated with the key concept and ask them to draw what it meant. I found that having students visualize and draw a picture to go along with their definition of it helped them make sense of the concept.

I monitored student progress in three main ways.

- First, as a primary teacher, I didn't have time to always collect work samples or to sit down with each student and talk about their learning. So what I did was schedule times for assessing their completed work. This could be one collection of completed work per student per week.

- Second, for conversations, I had a conference with each student at least one time every two weeks. Some students needed more than this. I had to remember that equal isn't fair.
- But what I used daily was observation. I would look at how my students were engaging with the concept and working with others.

Through this three-pronged approach to assessment, I had a good idea of who was doing what. It allowed me to see student learning in different contexts, which helped me see if they were generalizing the concept and able to apply it to unfamiliar problems.

### Success using a daily cumulative review

I was spending so much time on re-teaching during the year. I had to reteach multiplication of two-digit by two-digit whole numbers, division involving decimals, reading numbers, and comparing and ordering fractions, among other things. It got to the point that I wasn't able to address all the curriculum in the space of a school year.

I decided that I should start doing some sort of daily cumulative review. I picked four concepts to focus on: fractions, multiplication, division, and place value (reading and identifying the value of digits within the number). Each day, students saw one question for each concept.

By the time I started daily cumulative review, it was already November. At first, I was worried about making the questions, but the questions already exist in resources that we use. Plus, the questions aren't meant to be taxing, so it wasn't really a big process to come up with some.

It didn't take long. Students were soon able to remember these four concepts well. They could start using them right away when we worked on other concepts during the year.

I was surprised to see the impact on my year plan. Because I didn't have to reteach the concepts all the time, we were working through many other topics much more quickly than in the past. Students were making connections and could see when and how to apply the four concepts addressed during daily cumulative review.

One of the biggest indicators of the impact of this strategy came the following year. Another teacher shared how her students—mine from the previous year—were doing so well. They remembered many things from my grade and could apply these things to their current grade level's materials. They were remembering. Success!

## Success implementing a spiralled year plan and carry-over tests

I had only been teaching for a few years but wanted to approach my instruction a little differently. Though I wasn't ready to make big changes, I still thought it was time to support my students in their efforts to remember math content.

So I began by developing a different year plan. What I did was chunk each unit into three parts. I taught the introductory part of each unit during the first cycle, the middle part during the next cycle, and the third part during the final cycle. Each cycle had students digging deeper into the topic.

Initially, students didn't like leaving topics so soon. But when we returned to the concepts during the next cycle, they were able to make connections to other topics and strengthen their understanding of the topic.

To go along with this, I wanted to change testing. It didn't feel right to move on from things if students were struggling with them. The spiralled year plan helped, but I needed to do more. So I used carry-over tests. If I gave a test and noticed that many students struggled with a particular question or concept, I would add it to the next test, even long after I'd finished teaching that concept. I wanted to address the misconceptions students had instead of leaving them behind. After each test, I provided students with feedback about concepts that they struggled with. I gave them opportunities to strengthen their understanding through follow-up questions on future tests.

What I noticed, over the year, was that students started to get a better grasp of what they knew and didn't know. It was removing self-doubt from them, which made them more independent in the classroom.

## Success in teaching for learning instead of the quiz

I've seen lots of instructional approaches come down the pipes over the years. Every time, things don't seem to change in terms of student learning. What does change is teachers having additional pressures put on them. I've seen so many teachers be directed to completely change their approach. Definitely causes lots of stress.

This approach is different, and it makes a lot of sense. It isn't about starting from scratch. Instead, it's about being purposeful in planning instruction. Teachers can still use many of the strategies they have and just refine how they use these strategies. They need to think about spacing out teaching of concepts throughout the year and mixing concepts so that students have to pause and consider which strategy to apply.

I thought it would be impactful, but not as much as it actually was. Students were remembering concepts from earlier in the year and some from previous years. We weren't used to this. It was a realization that we are now teaching for learning instead of teaching for students to do well on a quiz the following day. In the approach we are using now, students are doing well on the test and continuing to do well. There is no more cramming in math class before provincial assessments. There is no need to. During the year, teachers are pulling concepts to the forefront in a way that makes sense and isn't overwhelming.

Our school scores have improved, and teachers are seeing how they can keep working with this approach. It is something that any teacher can see themselves doing. Teachers are all in different spots on the instructional continuum and this approach to instruction and learning gives everyone a makeable first step.

## Moving Forward

Regardless of where you are in your journey as a teacher, I hope that this book can be a support. Perhaps it can help you take your first step towards teaching math for learning math. Perhaps it can be a support to take you to the next level. During your journey, you may experience a paradigm shift in what matters in your math classroom. Perhaps your focus will shift from giving scores on quizzes and tests to determining if your students have an accurate understanding of their learning and are able to recall and apply previous learning to new situations.

The student learning strategies and instructional strategies we have explored in this book can work individually or in combination. You can pick and choose what helps you to craft a learning experience that promotes learning in your particular classroom. The journey can be hard, but it is necessary if you want to meet the instructional goal of helping your students become independent learners. Remember: it's about making math *stick*!

The journey can be hard, but it is necessary if you want to meet the instructional goal of helping your students become independent learners. Remember: it's about making math *stick*!

# References

Agarwal, P. K., & Bain, P. M. (2019). *Powerful teaching: Unleash the science of learning*. San Francisco: Jossey-Bass.

Boaler, J. (2016). *Mathematical mindsets: Unleashing students' potential through creative math, inspiring messages and innovative teaching*. Chappaqua: Jossey-Bass/Wiley.

Brown, P. C., Roediger, H. L. III, & McDaniel, M. A. (2014). *Make it stick: The science of successful learning*. Cambridge: The Belknap Press of Harvard University Press.

Clay, M. M. (1991). *Becoming literate: The construction of inner control*. Portsmouth: Heinemann.

Davies, A. (2011). *Making classroom assessment work* (3rd ed.). Courtenay: Connections Publishing.

Dweck, C. S. (2006). *Mindset: The new psychology of success*. New York: Ballantine Books.

Hattie, J. (2009). *Visible learning: A synthesis of over 800 meta-analyses relating to achievement*. New York: Routledge.

Karpicke, J. D. (2009). Metacognitive control and strategy selection: Deciding to practice retrieval during learning. *Journal of Experimental Psychology: General, 138*(4), 469–486.

Karpicke, J. D., & Blunt, J. R. (2011). Retrieval practice produces more learning than elaborative studying with concept mapping. *Science, 331*(6018), 772–775.

Leinwand, S. (2009). *Accessible mathematics: Ten instructional shifts that raise student achievement*. Portsmouth: Heinemann.

Leinwand, S. (2017, April). *10 Instructional tweaks every mathematics leader needs to advocate for and be able to model*. Paper presented at the 49th National Council of Supervisors of Mathematics Annual Conference, San Antonio, Texas.

National Council of Teachers of Mathematics. (2000). *Principles and standards for school mathematics*. Reston: NCTM.

Nazari, K. B., & Ebersbach, M. (2019). Distributed practice in mathematics: Recommendable especially for students on a medium performance level? *Trends in Neuroscience and Education, 17*, Article 100122. https://doi.org/10.1016/j.tine.2019.100122

Novak, J. D. (2013). Concept mapping. In J. Hattie & E. M. Anderman (Eds.), *International Guide to Student Achievement* (pp. 362–365). New York: Routledge/ Taylor & Francis Group.

Paivio, A. (1986). *Mental representations: A dual coding approach*. New York: Oxford University Press.

Park, J., & Brannon, E. M. (2013). Training the approximate number system improves math proficiency. *Psychological Science, 24*(10), 2013–2019.

Roediger, H. L. III, & Karpicke, J. D. (2006). The power of testing memory: Basic research and implications for educational practice. *Perspectives on Psychological Science, 1*(3), 181–210.

Roediger, H. L. III, & Karpicke, J. D. (2018). Reflections on the resurgence of interest in the testing effect. *Perspectives on Psychological Science, 13*(2), 236–241.

Rohrer, D. (2009). The effects of spacing and mixing practice problems. *Journal for Research in Mathematics Education, 40*(1), 4–17.

Scardamalia, M., & Bereiter, C. (1987). Knowledge telling and knowledge transforming in written composition. In S. Rosenberg (Ed.), *Advances in applied psycholinguistics, Vol. 2: Reading, writing and language learning* (pp. 142–175). New York: Cambridge University Press.

Schmidt, R. A., & Bjork, R. A. (1992). New conceptualizations of practice: Common principles in three paradigms suggest new concepts for training. *Psychological Science, 3*(4), 207–18.

Small, M. (2013). *Making math meaningful to Canadian students, K-8*. Toronto: Nelson Education Ltd.

Weinstein, Y., Sumeracki, M., & Caviglioli, O. (2019). *Understanding how we learn: A visual guide*. New York: Routledge.

# Recommended Resources

Boaler, J. (2016). *Mathematical mindsets: Unleashing students' potential through creative math, inspiring messages and innovative teaching*. Chappaqua: Jossey-Bass/Wiley.

Brown, P. C., Roediger, H. L. III, & McDaniel, M. A. (2014). *Make it stick: The science of successful learning*. Cambridge: The Belknap Press of Harvard University Press.

Costello, D. (2019). *Using what works: Strategies for developing a literacy-rich environment in math*. Oakville: Rubicon Publishing Inc.

Davies, A. (2011). *Making classroom assessment work*, (3rd ed.). Courtenay: Connections Publishing.

Dweck, C. S. (2006). *Mindset: The new psychology of success*. New York, NY: Ballantine Books.

Hattie, J. (2009). *Visible learning: A synthesis of over 800 meta-analyses relating to achievement*. New York: Routledge.

Humphreys, C., & Parker, R. (2015). *Making number talks matter*. Portland; Stenhouse

Kazemi, E., & Hintz, A. (2014). *Intentional talk: How to structure and lead productive mathematical discussions*. Portland: Stenhouse.

Krpan, C. M. (2017). *Teaching math with meaning: Cultivating self-efficacy through learning competencies, grades K–8*. Toronto: Pearson Canada.

Lawson, A. (2015). *What to look for: Understanding and developing student thinking in early numeracy*. Toronto: Pearson Canada.

Leinwand, S. (2009). *Accessible mathematics: Ten instructional shifts that raise student achievement*. Portsmouth: Heinemann.

National Council of Teachers of Mathematics. (2000). *Principles and standards for school mathematics*. Reston: National Council of Teachers of Mathematics.

National Council of Teachers of Mathematics. (2014). *Principles to actions: Ensuring mathematical success for all*. Reston: National Council of Teachers of Mathematics.

Parker, R., & Humphreys, C. (2018). *Digging deeper*. Portland: Stenhouse.

Small, M. (2009). *Big ideas from Dr. Small: Creating a comfort zone for teaching mathematics, grades 4–8*. Toronto: Nelson Education.

Small, M. (2010). *Big ideas from Dr. Small: Creating a comfort zone for teaching mathematics, grades K–3*. Toronto: Nelson Education.

Small, M. (2012). *Good questions: Great ways to differentiate mathematics instruction*. New York, NY: Teachers College Press.

Small, M. (2013). *Making math meaningful to Canadian students, K-8*. Toronto: Nelson Education Ltd.

Van de Walle, J. A., Lovin, L. H., Karp, K. S., & Bay-Williams, J. M. (2014). *Teaching student-centred mathematics: Developmentally appropriate instruction for grades pre-K-2* (2nd. Ed., Vol. 1). Upper Saddle River: Pearson Education, Inc.

Van de Walle, J. A., Lovin, L. H., Karp, K. S., & Bay-Williams, J. M. (2014). *Teaching student-centred mathematics: Developmentally appropriate instruction for grades 3-5* (2nd. Ed., Vol. 2). Upper Saddle River: Pearson Education, Inc.

Van de Walle, J. A., Lovin, L. H., Karp, K. S., & Bay-Williams, J. M. (2014). *Teaching student-centred mathematics: Developmentally appropriate instruction for grades 6-8* (2nd. Ed., Vol. 3). Upper Saddle River: Pearson Education, Inc.

West, L. (2018). *Adding talk to the equation*. Portland: Stenhouse.

Zager, T. J. (2017). *Becoming the math teacher you wish you'd had*. Portland: Stenhouse.

# Index